W0106993

Lecture Notes in Physics

For information about Vols. 1–23, please contact your bookseller or Springer-Verlag.

Vol. 24: R. F. Snipes, Statistical Mechanical Theory of the Electrolytic Transport of Nonelectrolytes. V, 210 pages. 1973.

Vol. 25: Constructive Quantum Field Theory. The 1973 "Ettore Majorana" International School of Mathematical Physics. Edited by G. Velo and A. Wightman. III, 331 pages. 1973.

Vol. 26: A. Hubert, Theorie der Domänenwände in geordneten Medien. XII, 377 Seiten. 1974.

Vol. 27: R. K. Zeytounian, Notes sur les Ecoulements Rotationnels de Fluides Parfaits. XIII, 407 pages. 1974.

Vol. 28: Lectures in Statistical Physics. Edited by W. C. Schieve and J. S. Turner. V, 342 pages. 1974.

Vol. 29: Foundations of Quantum Mechanics and Ordered Linear Spaces. Advanced Study Institute, Marburg 1973. Edited by A. Hartkämper and H. Neumann. VI, 355 pages. 1974.

Vol. 30: Polarization Nuclear Physics. Proceedings 1973. Edited by D. Fick. IX, 292 pages. 1974.

Vol. 31: Transport Phenomena. Sitges International Schools of Statistical Mechanics, June 1974. Edited by G. Kirczenow and J. Marro. XIV, 517 pages. 1974.

Vol. 32: Particles, Quantum Fields and Statistical Mechanics. Proceedings 1973. Edited by M. Alexanian and A. Zepeda. V, 132 pages. 1975.

Vol. 33: Classical and Quantum Mechanical Aspects of Heavy Ion Collisions. Proceedings 1974. Edited by H. L. Harney, P. Braun-Munzinger, and C. K. Gelbke. VII, 311 pages. 1975.

Vol. 34: One-Dimensional Conductors GPS Summer School Proceedings, 1974. Edited by H. G. Schuster. VII, 371 pages. 1975.

Vol. 35: Proceedings of the Fourth International Conference on Numerical Methods in Fluid Dynamics, 1974. Edited by R. D. Richtmyer. V, 457 pages. 1975.

Vol. 36: R. Gatignol, Théorie Cinétique des Gaz à Répartition Discrète de Vitesses. II, 219 pages. 1975.

Vol. 37: Trends in Elementary Particle Theory. Proceedings 1974. Edited by H. Rollnik and K. Dietz. V, 472 pages. 1975.

Vol. 38: Dynamical Systems, Theory and Applications. Proceedings 1974. Edited by J. Moser. VI, 624 pages. 1975.

Vol. 39: International Symposium on Mathematical Problems in Theoretical Physics. Proceedings 1975. Edited by H. Araki. XII, 562 pages. 1975.

Vol. 40: Effective Interactions and Operators in Nuclei. Proceedings 1975. Edited by B. R. Barrett. XII, 339 pages. 1975.

Vol. 41: Progress in Numerical Fluid Dynamics. Proceedings 1974. Edited by H. J. Wirz. V, 471 pages. 1975.

Vol. 42: H II Regions and Related Topics. Proceedings 1975. Edited by D. Downes and T. L. Wilson. XII, 488 pages. 1975.

Vol. 43: Laser Spectroscopy. Proceedings 1975. Edited by S. Haroche, J. C. Pebay-Peyroula, T. W. Hänsch, and S. E. Harris. X, 466 pages. 1975.

Vol. 44: R. A. Breuer, Gravitational Perturbation Theory and Synchrotron Radiation. VI, 196 pages. 1975.

Vol. 45: Dynamical Concepts on Scaling Violation and the New Resonances in e^+e^- Annihilation. Edited by B. Humpert. VII, 248 pages. 1976.

Vol. 46: E. J. Flaherty, Hermitian and Kählerian Geometry in Relativity. VIII, 365 pages. 1976.

Vol. 47: Padé Approximants Method and Its Applications to Mechanics. Edited by H. Cabannes. XV, 267 pages. 1976.

Vol. 48: Interplanetary Dust and Zodiacal Light. Proceedings 1975. Edited by H. Elsässer and H. Fechtig. XII, 496 pages. 1976.

Vol. 49: W. G. Harter and C. W. Patterson, A Unitary Calculus for Electronic Orbitals. XII, 144 pages. 1976.

Vol. 50: Group Theoretical Methods in Physics. 4th International Colloquium. Nijmegen 1975. Edited by A. Janner, T. Janssen, and M. Boon. XIII, 629 pages. 1976.

Vol. 51: W. Nörenberg und H. A. Weidenmüller. Introduction to the Theory of Heavy-Ion Collisions. IX, 273 pages. 1976.

Vol. 52: M. Mladjenović, Development of Magnetic β-Ray Spectroscopy. X, 282 pages. 1976.

Vol. 53: D. J. Simms and N. M. J. Woodhouse, Lectures on Geometric Quantization. V, 166 pages. 1976.

Vol. 54: Critical Phenomena. Sitges International School on Statistical Mechanics, June 1976. Edited by J. Brey and R. B. Jones. XI, 383 pages. 1976.

Vol. 55: Nuclear Optical Model Potential. Proceedings 1976. Edited by S. Boffi and G. Passatore. VI, 221 pages. 1976.

Vol. 56: Current Induced Reactions. International Summer Institute, Hamburg 1975. Edited by J. G. Körner, G. Kramer, and D. Schildknecht. V, 553 pages. 1976.

Vol. 57: Physics of Highly Excited States in Solids. Proceedings 1975. Edited by M. Ueta and Y. Nishina. IX, 391 pages. 1976.

Vol. 58: Computing Methods in Applied Sciences. Proceedings 1975. Edited by R. Glowinski and J. L. Lions. VIII, 593 pages. 1976.

Vol. 59: Proceedings of the Fifth International Conference on Numerical Methods in Fluid Dynamics. 1976. Edited by A. I. van de Vooren and P. J. Zandbergen. VII, 459 pages. 1976.

Vol. 60: C. Gruber, A. Hintermann, and D. Merlini, Group Analysis of Classical Lattice Systems. XIV, 326 pages. 1977.

Vol. 61: International School on Electro and Photonuclear Reactions I. Edited by C. Schaerf. VIII, 650 pages. 1977.

Vol. 62: International School on Electro and Photonuclear Reactions II. Edited by C. Schaerf. VIII, 301 pages. 1977.

Vol. 63: V. K. Dobrev et al., Harmonic Analysis on the n-Dimensional Lorentz Group and Its Application to Conformal Quantum Field Theory. X, 280 pages. 1977.

Vol. 64: Waves on Water of Variable Depth. Edited by D. G. Provis and R. Radok. 231 pages. 1977.

Vol. 65: Organic Conductors and Semiconductors. Proceedings 1976. Edited by L. Pál, G. Grüner, A. Jánossy and J. Sólyom. 654 pages. 1977.

Vol. 66: A. H. Völkel, Fields, Particles and Currents. VI, 354 pages. 1977.

Vol. 67: W. Drechsler and M. E. Mayer, Fiber Bundle Techniques in Gauge Theories. X, 248 pages. 1977.

Lecture Notes in Physics

Edited by J. Ehlers, München, K. Hepp, Zürich
R. Kippenhahn, München, H. A. Weidenmüller, Heidelberg
and J. Zittartz, Köln
Managing Editor: W. Beiglböck, Heidelberg

97

L. P. Hughston

Twistors and Particles

Springer-Verlag
Berlin Heidelberg GmbH 1979

Author

Lane Palmer Hughston
The Mathematical Institute
University of Oxford
Oxford
England

ISBN 978-3-540-09244-5 ISBN 978-3-540-35336-2 (eBook)
DOI 10.1007/978-3-540-35336-2

Library of Congress Cataloging in Publication Data. Hughston, L P 1951-
Twistors and particles. (Lecture notes in physics ; 97) Bibliography: p. Includes index.
1. Particles (Nuclear physics) 2. Twistor theory. I. Title. II. Series.
QC793.3.F5H83 539.7'21 79-13891

This work is subject to copyright. All rights are reserved, whether the whole or part of the material is concerned, specifically those of translation, reprinting, re-use of illustrations, broadcasting, reproduction by photocopying machine or similar means, and storage in data banks. Under § 54 of the German Copyright Law where copies are made for other than private use, a fee is payable to the publisher, the amount of the fee to be determined by agreement with the publisher.

© by Springer-Verlag Berlin Heidelberg 1979

Originally published by Springer-Verlag Berlin Heidelberg New York in 1979

2153/3140-543210

PREFACE

The momentum of the mind is all toward abstraction.

- Wallace Stevens, Opus Posthumous

Within the framework of twistor theory the structure of spacetime is relegated, in contrast to the position which it has held since the beginning of the twentieth century, to a status of secondary character. Whereas in the past spacetime has always served as the background against which phenomena are to be interpreted—and indeed, according to Einstein's theory of gravitation, spacetime serves moreover as a basic dynamical entity itself—the new view which the twistor theorists are advocating takes twistor space, with the many rich and variegated aspects of its complex analytic structure, as the primary descriptive device and dynamical construction in terms of which phenomena are to be understood.

The difficulties inherent in a spacetime description have long been appreciated by many authors. Julian Schwinger, for example, in his preface to Selected Papers on Quantum Electrodynamics summarizes the situation aptly when he remarks that "... The localization of charge with indefinite precision requires for its realization a coupling with the electromagnetic field that can obtain arbitrarily large magnitudes. The resulting appearance of divergences, and contradictions, serves to deny the basic measurement hypothesis. We conclude that a convergent theory cannot be formulated consistently within the framework of present space-time concepts. To limit the magnitude of interactions while retaining the customary coordinate description is contradictory, since no mechanism is provided for precisely localized measurements." With a similar attitude towards this question Einstein, at the end of The Meaning of Relativity, concludes that "One can give good reasons why reality cannot at all be represented by a continuous field. From the quantum phenomena it appears to follow with certainty that a finite system of finite energy can be completely described by a finite set of numbers (quantum numbers). This does not seem to be in accordance with a continuum theory, and must lead to an attempt to find a purely algebraic theory for the description of reality." Of course when he refers to a continuum Einstein means spacetime, taken with its usual real differentiable structure. In twistor theory, however, the continuum which arises is that of the complex number system, and those aspects of the geometry of twistor space which are of interest to physics stem more specifically from its complex analytic structure, rather than its real differentiable structure. The general characterization of the structures which can arise in the case of complex analytic manifolds has been the subject of intense investigation by mathematicians, especially with the advent of the powerful techniques of sheaf cohomology theory. One of the precepts of twistor theory is that here, within a suitably formulated sheaf cohomological framework, we have the proper basis for a "purely algebraic" description that is compatible both with the ideas of relativity and with the principles of quantum mechanics.

This view has met with a reasonable degree of success, and it has been possible, using methods of algebraic geometry and complex analytic geometry, for twistor theorists to assemble the outlines of a new approach to elementary particle physics. The subject is still in its infancy and in a rapid state of development, and thus many of its results are only of a preliminary character and are both subject to and deserving of considerable modification and improvement. In spite of their tentative nature, it seemed appropriate nonetheless to prepare an account of some of these matters for a wider audience, with the hope that it might stimulate or otherwise prove a useful aid in further and more extensive research into the subject. With this purpose in mind the following study is presented.

Although a fair amount of background material is covered in Chapters 2 and 3, the reader previously uninitiated into the mysteries of twistor theory may find it necessary to consult some additional references. For the two-component spinor formalism see Pirani (1965), Penrose (1968a), and the forthcoming book by Rindler and Penrose. For further reading in basic twistor theory see Penrose (1967), Penrose and MacCallum (1972), and Penrose (1975a). Although a specialized knowledge of elementary particle physics is not necessary, at the outset, for reading this volume, it is assumed nonetheless that the reader is familiar with basic quantum mechanics, and is acquainted already, to some extent, with the quark model.

The author is indebted to many of his colleagues for their help in the preparation and development of this material, particularly to R. Penrose who originated many of the ideas discussed here, and who has acted as a constant source of illumination and inspiration. G.A.J. Sparling has contributed extensively to this work, and the author wishes to thank him for many helpful discussions. I would also like to thank many of my colleagues at Oxford and elsewhere, including D.M. Blasius, M. Eastwood, M.L. Ginsberg, A. Hodges, S.A. Huggett, T.R. Hurd, R. Jozsa, E.T. Newman, A. Popovich, Z. Perjés, I. Robinson, M. Sheppard, L. Smarr, P. Sommers, K.P. Tod, Tsou S.T., M. Walker, R.S. Ward, and N.M.J. Woodhouse, for useful conversations and suggestions related to the work described herein. The author is grateful to B.S. DeWitt, C.M. DeWitt, R. Matzner, L. Shepley, H.J. Smith, and the late Alfred Schild, as well as other colleagues at the University of Texas at Austin, for their hospitality shown during the author's 1974 visit, when some of the ideas preliminary to the material described here were worked out. The author has profited much from his regular visits, supported by the Clark Foundation, to the University of Texas at Dallas, and he would like to thank I. Ozsvath, W. Rindler, I. Robinson, and J.R. Robinson for their hospitality. Likewise the author has benefited from his visits to the Astronomy Department at the University of Virginia, and gratitude is expressed to W. Saslaw, and other colleagues there, for their hospitality. I am grateful to J. Ehlers, M.L. Ginsberg, C.J. Isham, R. Penrose, G.A.J. Sparling, and N.M.J. Woodhouse for reading earlier drafts of the manuscript and contributing many corrections and helpful suggestions for improvement.

This work was supported by a Rhodes Scholarship at Oxford during the years 1972-1975. This work was also supported by the Westinghouse Corporation during 1972-73. More recently the work described herein has been supported by a grant from the Science Research Council, and by a Junior Research Fellowship at Wolfson College, Oxford. I am very grateful to Valerie Censabella, who typed the manuscript and who has been most helpful at all stages in the preparation of this material.
This volume is dedicated to my mother and my father.

TABLE OF CONTENTS

Page

CHAPTER 1

INTRODUCTORY REMARKS

Progress in any aspect is a movement
through changes in terminology.

- Wallace Stevens, Opus Posthumous

This study will touch on a variety of topics concerning twistor theory and elementary particle physics. A few of these topics will be treated in some detail, but none exhaustively. The purpose of this work is to describe how it is possible, using twistor methods, to gain some understanding of the microscopic structural degrees of freedom responsible for the properties of elementary particles.

In a very general sense the methodology of twistor theory consists simply of the application of techniques of complex analytic geometry to problems in physics. Inherent in the twistor program are many changes in terminology, whereby a number of the familiar concepts of physics are reexpressed in the language of algebraic geometry and analytic geometry. "The physicist always prefers to sacrifice the less perfect concepts of physics rather than the simpler, more perfect, and more lasting concepts of geometry, which form the solidest foundation of all his theories", said Mach, and there is certainly a good deal of reason in his remark: but the twistor philosophy goes one step further, and insists that within geometry itself one can discover all the laws of physics.

The organization of this volume is as follows. Chapters 2 and 3 view twistor space from the standpoint of classical physics. Algebraic geometry is to complex analytic geometry as classical physics is to quantum physics—and in Chapters 2 and 3 twistor space is explored with various tools of algebraic geometry. Most of the information in Chapter 2 is standard background material, and is summarized here for the reader previously unacquainted with twistor theory. Twistors are first defined in terms of classical systems of zero rest mass—that is to say, classical special relativistic systems defined by a null momentum and an angular momentum which is related to the momentum in such a way that twistors transform in a natural way under the action of the group SU(2,2), and, in particular, the Poincaré group. In §2.4 it is shown that twistors can be characterized in terms of the solutions of a certain differential equation called the "twistor equation". In §§2.5 and 2.6 twistors are described in terms of the geometry of complex projective 3-space P^3. Complex projective lines in P^3 correspond to points in complex Minkowski space; using this correspondence (the "Klein representation") various aspects of the geometry of spacetime are expressed in twistor terms, and vice-versa.

In Chapter 3 it is shown how massive systems can be built up out of two or more twistors. The momentum and the angular momentum are described in terms of a single two-index symmetric "kinematical twistor". Theorem 3.3.1 shows how any massive system can be decomposed into two or more twistor constituents. Thus, massive systems (at the classical level) can always be regarded as being "made up" out of twistors. Twistors are, in a certain sense, the elementary constituents of matter. For a given momentum and angular momentum there are internal degrees of freedom which yet remain, mixing the various twistor constituents. Theorems 3.4.2 and 3.4.14 show the relevant groups which leave the momentum and angular momentum of an n-twistor system invariant. These groups are called the "n-twistor internal symmetry groups", and, for each value of n, contain U(n) as a subgroup. It is proposed that these internal degrees of freedom are in some sense responsible for the phenomenological unitary groups which arise naturally in elementary particle classification schemes (e.g., SU(3)). In §3.5 a center of mass twistor is introduced for n-twistor systems. This construction plays a useful role in a number of problems.

In Chapter 4 the rules of twistor quantization are introduced for systems composed of a single twistor. It is shown how solutions of the zero rest mass equations can be obtained in terms of holomorphic functions defined over suitable domains of twistor space. Both positive and negative helicity fields are discussed, and the differences in the relevant contour integral formulae for evaluating the fields, in the two cases, are noted. The positive frequency condition is discussed in §4.5, and the whole procedure is illustrated with the example of an elementary state.

In Chapter 5 massive fields are described in terms of holomorphic functions of two or more twistors. It is proposed that observables correspond to holomorphic differential operators with polynomial coefficients. Explicit expressions are presented for the operators corresponding to momentum, angular momentum, mass, and spin. In §5.5 the operators corresponding to "internal" observables are discussed, and are described explicitly in the cases of one, two, and three twistors.

In Chapter 6 the scheme is applied to the low-lying baryons—that is to say, the N(949) octet and the Δ(1232) decimet. After a brief review of the quark model (described in a language suitable for our purposes) it is demonstrated how the low-lying baryons can be represented in terms of certain types of holomorphic functions of three twistors. Baryons are not regarded as bound states of quarks. No color degrees of freedom are introduced.

In Chapter 7 the methods of Chapter 6 are extended so as to apply to more general systems. Mesons are introduced as quark-antiquark bound states, described in terms of holomorphic functions of six twistors. The charge conjugation quantum number plays a crucial role in the representation of these states. Orbital angular momentum is described in twistor terms, and it is shown how orbital excitations of the quark-antiquark system lead to meson resonances. Baryon resonances are represented as excitations of a quark-diquark bound state. The deuteron is briefly discussed, from a twistor point of view, in the last section of Chapter 7. In Chapter 8, after a review of the properties of leptons and of parity violation in weak interactions, a model for sequential leptons is built up in twistor terms. Chapters 9 and 10 are concerned with further mathematical developments in the theory. In Chapter 9 the methods of sheaf cohomology are introduced, and these are applied to various problems in Chapter 10, the aim being to sharpen up much of the material of the previous chapters, and to open up the doors to more extensive developments.

The tentative nature of any general inferences that can now be put forward in connection with the twistor particle program, or, for that matter, twistor theory in general, should undoubtedly be apparent to anybody working in this subject. One need merely consider the vast range of phenomena which so far have resisted any formulation in twistor terms whatsoever. Nonetheless, significant conclusions are being drawn along certain lines, and are receiving continually increasing support.

In particular, the central role of the twistor program in connection with Einstein's theory seems to me now firmly established, and there does not seem to be any reason now why particle physics as a whole should not be amenable to treatment within the framework of twistor theory.

CHAPTER 2

ASPECTS OF THE GEOMETRY OF TWISTOR SPACE

2.1 Classical Systems of Zero Rest Mass.

There are various ways of building up the framework of twistor theory, and it must be said that it is not exactly clear where to begin. For the purposes of investigations into elementary particle physics a convenient, if not totally adequate, place to start is with the observation that a point Z^{α} in twistor space ($\alpha = 0,1,2,3$) can be represented naturally in terms of physical quantities as a classical system of zero rest mass.

Such a system is characterized by its total momentum P^a, which is null and future-pointing, and its angular momentum M^{ab} ($= -M^{ba}$) with respect to a particular choice of origin in spacetime.

Together these quantities must satisfy a relation to the effect that if we form the spin-vector

(2.1.1) $$S_a = \frac{1}{2}\,\varepsilon_{abcd}P^bM^{cd}$$

then the proportionality $S_a = sP_a$ holds for some value of the number s. The magnitude of s is the spin of the system, and s itself is called the helicity. Positive helicity systems are called right-handed, and negative helicity systems are called left-handed.

There is a certain algebraic characterization of the momentum and angular momentum that ensures that together they constitute a zero rest mass (henceforth abbreviated ZRM) system:

2.1.2 Proposition. A pair $\{P^a, M^{ab}\}$ represents a ZRM system if and only if there exists a pair of spinors $(\omega^A, \pi_{A'})$ such that

(2.1.3) $$P^a = \bar{\pi}^A\pi^{A'}$$

and

(2.1.4) $$M^{ab} = i\omega^{(A}\bar{\pi}^{B)}\varepsilon^{A'B'} - i\bar{\omega}^{(A'}\pi^{B')}\varepsilon^{AB} ,$$

where $\bar{\pi}^A$ is the complex conjugate of $\pi^{A'}$, and $\bar{\omega}^{A'}$ is the complex conjugate of ω^A.

<u>Proof</u>[1]. The existence of a spinor $\pi^{A'}$ such that equation (2.1.3) is satisfied is precisely the condition that P^a should be null and future-pointing.

The spin-relation $S^a = sP^a$ can be written $*M^{ab}P_b = sP^a$ where $*M^{ab} := \frac{1}{2}\varepsilon^{abcd}M_{cd}$ is the dual of M^{ab}. If we write

$$(2.1.5) \qquad *M^{ab} = -i\mu^{AB}\varepsilon^{A'B'} + i\bar{\mu}^{A'B'}\varepsilon^{AB} ,$$

where μ^{AB} is a symmetric spinor, then the spin-relation, using equation (2.1.3), reads

$$(2.1.6) \qquad -i\mu^{AB}\bar{\pi}_B\pi^{A'} + i\bar{\mu}^{A'B'}\pi_{B'}\bar{\pi}^A = s\bar{\pi}^A\pi^{A'} .$$

Contracting this relation with $\bar{\pi}_A$ yields $\mu^{AB}\bar{\pi}_A\bar{\pi}_B = 0$, which implies $-i\mu^{AB} = \omega^{(A}\bar{\pi}^{B)}$ for some choice of ω^A, the factor of $-i$ being included for later convenience. Finally, using the fact that equation (2.1.5) implies

$$(2.1.7) \qquad M^{ab} = \mu^{AB}\varepsilon^{A'B'} + \bar{\mu}^{A'B'}\varepsilon^{AB} ,$$

we deduce equation (2.1.4). □

The spinor pair $(\omega^A , \pi_{A'})$ completely determines the ZRM system, and defines a point Z^α in twistor space according to the scheme

$$(2.1.8) \qquad (Z^0 , Z^1 , Z^2 , Z^3) = (\omega^0 , \omega^1 , \pi_{0'} , \pi_{1'}) .$$

Note, on the other hand, that a ZRM system determines its associated twistor only up to an overall phase factor, since the momentum and the angular momentum are invariant under the transformation

$$(2.1.9) \qquad (\omega^A , \pi_{A'}) \longrightarrow e^{i\theta}(\omega^A , \pi_{A'}) .$$

It is interesting to observe that the helicity of a ZRM system can be expressed directly in twistor terms. For this purpose it is useful to define the complex conjugate twistor $\bar{Z}_\alpha$ by the spinor pair $(\bar{\pi}_A , \bar{\omega}^{A'})$. A short calculation establishes that the inner product defined by

$$(2.1.10) \qquad Z^\alpha\bar{Z}_\alpha = \omega^A\bar{\pi}_A + \pi_{A'}\bar{\omega}^{A'}$$

is precisely twice the helicity of the system, i.e. we have $Z^{\alpha}\bar{Z}_{\alpha} = 2s$.

One might be inclined initially to think that the freedom expressed in (2.1.9) is of an irrelevant nature, and arises perhaps on account of some slight inadequacy in the representation that has been chosen for twistors in terms of systems of zero rest mass. Nothing could be further from the truth, however. One of the remarkable things about twistors is that they do, in fact, carry more information in them than just momentum and angular momentum. This fact takes on great significance, as we shall see, when quantum mechanics is brought into the picture.

2.2 The Action of the Poincaré Group.

It is of considerable interest to know how the action of the Poincaré group is expressed in twistor terms. Since our ultimate goal is to express various field quantities in terms of twistors, and since these field quantities must themselves be subject to a particular behavior under the action of the Poincaré group, it is of significance to study the action of the Poincaré group on twistors first.

Under the spacetime translation $x^a \longrightarrow x^a + r^a$ the angular momentum M^{ab} transforms according to the rule

(2.2.1) $$M^{ab} \longrightarrow M^{ab} + 2r^{[a}P^{b]} \quad .$$

It is not difficult to check that for a ZRM system the transformation on Z^{α} that induces (2.2.1) is

(2.2.2) $$\omega^{A} \longrightarrow \omega^{A} + ir^{AA'}\pi_{A'} \quad , \quad \pi_{A'} \longrightarrow \pi_{A'} \quad .$$

This transformation can therefore be regarded as defining the action of a spacetime translation on Z^{α}.

The action of a restricted Lorentz transformation on a ZRM system is specified by

(2.2.3) $$P_a \longrightarrow \Lambda_a{}^{b}P_b \quad , \quad M_{ab} \longrightarrow \Lambda_a{}^{c}\Lambda_b{}^{d}M_{cd} \quad .$$

For a restricted Lorentz transformation $\Lambda_a{}^{b}$ has the form

(2.2.4) $$\Lambda_a{}^{b} = \ell_A{}^{B}\bar{\ell}_{A'}{}^{B'} \quad ,$$

where $\ell_A{}^B$ is an element of the group SL(2,C), i.e. subject to the relation

$$\ell_A{}^C \ell_B{}^D \varepsilon_{CD} = \varepsilon_{AB} \quad . \tag{2.2.5}$$

The action on Z^α which induces (2.2.3) is easily verified to be:

$$\omega^A \longrightarrow -\ell^A{}_B \omega^B \quad , \quad \pi_{A'} \longrightarrow \bar{\ell}_{A'}{}^{B'} \pi_{B'} \quad . \tag{2.2.6}$$

By following a Lorentz transformation with a translation, we can realize the complete action of the restricted Poincaré group on a twistor. This can be conveniently expressed in the form

$$Z^\alpha \longrightarrow P^\alpha{}_\beta Z^\beta \quad , \tag{2.2.7}$$

where the transformation matrix $P^\alpha{}_\beta$ is given by

$$P^\alpha{}_\beta = \begin{pmatrix} -\ell^A{}_B & ir^{AA'}\bar{\ell}_{A'}{}^{B'} \\ 0 & \bar{\ell}_{A'}{}^{B'} \end{pmatrix} \quad , \tag{2.2.8}$$

with the usual laws of matrix multiplication applying in the contraction of $P^\alpha{}_\beta$ with the spinor parts of Z^β. That is to say, we have

$$\omega^A \longrightarrow -\ell^A{}_B \omega^B + ir^{AA'}\bar{\ell}_{A'}{}^{B'}\pi_{B'} \qquad \pi_{A'} \longrightarrow \bar{\ell}_{A'}{}^{B'}\pi_{B'} \tag{2.2.9}$$

for the spinor parts of equation (2.2.7).

2.3 The Group SU(2,2).

The complex conjugate twistor $\bar{Z}_\alpha$ undergoes the complex conjugate transformation $\bar{Z}_\alpha \longrightarrow \bar{P}_\alpha{}^\beta \bar{Z}_\beta$ when Z^α undergoes transformation (2.2.7). Since the helicity s is Poincaré invariant, the requirement that the inner product $Z^\alpha \bar{Z}_\alpha$ be preserved implies that $P^\gamma{}_\beta \bar{P}_\gamma{}^\alpha = \delta^\alpha_\beta$, where δ^α_β is the twistor Kronecker delta, given in spinor parts by:

$$\delta^\alpha_\beta = \begin{pmatrix} -\varepsilon^A{}_B & 0 \\ 0 & \varepsilon_{A'}{}^{B'} \end{pmatrix} \quad . \tag{2.3.1}$$

The set of all matrices $U^\alpha{}_\beta$ satisfying $U^\gamma{}_\beta \bar{U}_\gamma{}^\alpha = \delta^\alpha_\beta$ forms the group U(2,2). This can be seen as follows. Such transformation matrices preserve the norm $Z^{\alpha}\bar{Z}_{\alpha}$, which is given explicitly by

$$Z^{\alpha}\bar{Z}_{\alpha} = \omega^{A}\bar{\pi}_{A} + \pi_{A'}\bar{\omega}^{A'} = \omega^{0}\bar{\pi}_{0} + \omega^{1}\bar{\pi}_{1} + \pi_{0'}\bar{\omega}^{0'} + \pi_{1'}\bar{\omega}^{1'} \quad . \tag{2.3.2}$$

If new variables are introduced according to the scheme

$$\omega^{0} = (w+y) \qquad \omega^{1} = (x+z) \qquad \pi_{0'} = (w-y) \qquad \pi_{1'} = (x-z) \tag{2.3.3}$$

where w, x, y, and z are complex, then

$$\tfrac{1}{2}Z^{\alpha}\bar{Z}_{\alpha} = w\bar{w} + x\bar{x} - y\bar{y} - z\bar{z} \quad , \tag{2.3.4}$$

which shows that the helicity is a quadratic Hermitian form of signature {++--}. The group U(2,2) is by definition the multiplicative group of complex linear transformations which preserve a quadratic Hermitian form of that signature.

The group SU(2,2) is the subgroup of U(2,2) consisting of matrices which, in addition to satisfying $U^{\gamma}{}_{\beta}\bar{U}_{\gamma}{}^{\alpha} = \delta^{\alpha}_{\beta}$, also preserve the twistor epsilon tensor $\varepsilon^{\alpha\beta\gamma\delta}$, i.e.:

$$U^{\alpha}{}_{\xi}U^{\beta}{}_{\eta}U^{\gamma}{}_{\zeta}U^{\delta}{}_{\theta}\,\varepsilon^{\xi\eta\zeta\theta} = \varepsilon^{\alpha\beta\gamma\delta} \quad . \tag{2.3.5}$$

Condition (2.3.5) amounts to the same thing as requiring that $U^{\alpha}{}_{\beta}$ have unit determinant.

SU(2,2) is of special importance to physics inasmuch as it is locally isomorphic with the 15-parameter conformal group of compactified Minkowski space[2]. The restricted Poincaré group is a subgroup of SU(2,2). A description of the relationship between the two groups can be facilitated with the introduction of the so-called "infinity twistors", given by

$$I^{\alpha\beta} = \begin{pmatrix} \varepsilon^{AB} & 0 \\ 0 & 0 \end{pmatrix} \qquad I_{\alpha\beta} = \begin{pmatrix} 0 & 0 \\ 0 & \varepsilon^{A'B'} \end{pmatrix} \tag{2.3.6}$$

which, according to a scheme to be elaborated in Section 2.6, represent the vertex of the null cone at infinity.

The infinity twistors are skew-symmetric, are complex conjugates of one-another, and satisfy the following relations:

$$I^{\alpha\beta}I_{\beta\gamma} = 0 \ , \quad I^{\alpha\beta} = \frac{1}{2}\varepsilon^{\alpha\beta\gamma\delta}I_{\gamma\delta} \ , \quad I_{\alpha\beta} = \frac{1}{2}\varepsilon_{\alpha\beta\gamma\delta}I^{\gamma\delta} \ . \tag{2.3.7}$$

Poincaré transformations are SU(2,2) transformations which have the property that they preserve the infinity twistors.

2.4 The Twistor Equation.

Another way in which twistor space arises naturally is as the solution set of the differential equation

$$\nabla^{A'(A}\xi^{B)} = 0 \ , \tag{2.4.1}$$

which, accordingly, is sometimes called the twistor equation.

2.4.2 Proposition. The general solution of equation (2.4.1) is

$$\xi^{A}(x) = \omega^{A} - ix^{AA'}\pi_{A'} \ , \tag{2.4.3}$$

where ω^A and $\pi_{A'}$ are constant.

Proof. Equation (2.4.1) can be written in the form

$$\nabla^{B'B}\xi^{C} = \frac{1}{2}\varepsilon^{BC}\nabla^{B'}_{D}\xi^{D} \ . \tag{2.4.4}$$

Taking a derivative, we have

$$\nabla^{A'A}\nabla^{B'B}\xi^{C} = \frac{1}{2}\varepsilon^{BC}\nabla^{AA'}\nabla^{B'}_{D}\xi^{D} \ , \tag{2.4.5}$$

which, using $\nabla^{A'(A}\xi^{C)} = 0$, implies

$$\varepsilon^{B(C}\nabla^{A)A'}\nabla^{B'}_{D}\xi^{D} = 0 \ , \tag{2.4.6}$$

showing that $\nabla^{B'}_{D}\xi^{D}$ is a constant spinor, which will be denoted $2i\pi^{B'}$, the factor of 2i being for convenience. Substituting this result back into equation (2.4.4), integration then gives (2.4.3), with ω^A appearing as a constant of integration. □

The pair $(\omega^A, \pi_{A'})$ defines the twistor Z^α, and $\xi^A(x)$ is called the associated spinor field(3) of the twistor Z^α. It can be checked that the natural action of the

Poincaré group on $\xi^A(x)$ agrees with the action on Z^α defined in Section 2.2.

2.5 α-Planes and β-Planes.

The location of a twistor Z^α in complex Minkowski space can be defined as the region for which the associated spinor field $\xi^A(x)$ vanishes. From (2.4.3) this is evidently the condition that

$$\omega^A = ix^{AA'}\pi_{A'} \quad . \tag{2.5.1}$$

Since equation (2.5.1) is linear in $x^{AA'}$, and represents a pair of conditions that these coordinates must satisfy, the solution for fixed ω^A and $\pi_{A'}$ must be a 2-plane. Moreover it should be obvious that if $x_0^{AA'}$ represents any particular point satisfying (2.5.1), then the general point satisfying this relation is $x_0^{AA'} + \lambda^A\pi^{A'}$, where the spinor λ^A is arbitrary. So the location of the twistor Z^α is the 2-plane consisting of all the endpoint positions of a complex vector $\lambda^A\pi^{A'}$ springing from the point $x_0^{AA'}$. Each such complex vector is null. Moreover, since $\pi^{A'}$ is fixed, each such vector is orthogonal to any other. Thus Z^α corresponds to a null 2-plane in Minkowski space.

A point W_α in <u>dual</u> twistor space is represented by a spinor pair $(\sigma_A , \tau^{A'})$. Associated with W_α is a solution of the "primed" twistor equation

$$\nabla^{A(A'}\eta^{B')} = 0 \tag{2.5.2}$$

given by

$$\eta^{A'} = \tau^{A'} + ix^{A'A}\sigma_A \quad . \tag{2.5.3}$$

By analogy with Proposition (2.4.2) it is not difficult to see that equation (2.5.3) gives the general solution of (2.5.2). The locus of the dual twistor W_α is given by

$$\tau^{A'} = -ix^{A'A}\sigma_A \quad , \tag{2.5.4}$$

the region where $\eta^{A'}$ vanishes. In this case if $x_0^{A'A}$ represents any particular solution to (2.5.4) then the general solution is given by $x_0^{A'A} + \lambda^{A'}\sigma^A$.

It is of interest to note that in complex Minkowski space there are two distinct systems of null 2-planes. The so-called α-planes are those null 2-planes which correspond to twistors of valence $[{}^{1}_{0}]$, i.e. the Z^{α}-type twistors. The β-planes are those null 2-planes which correspond to twistors of valence $[{}^{0}_{1}]$, i.e. the W_{α}-type twistors.

Any two distinct α-planes have a unique intersection point in complex Minkowski space. If the corresponding twistors are denoted Z_1^{α} and Z_2^{α} then for an intersection point one must solve simultaneously the algebraic equations,

$$\omega_1^{A} = ix^{AA'}\pi_{1A'}, \qquad \omega_2^{A} = ix^{AA'}\pi_{2A'} \quad . \tag{2.5.5}$$

Assuming that $\pi_{1A'}$ is not proportional to $\pi_{2A'}$, the unique solution to these equations is given by the formula

$$ix^{AA'} = (\omega_1^{A}\pi_2^{A'} - \omega_2^{A}\pi_1^{A'})/(\pi_{1A'}\pi_2^{A'}) \quad , \tag{2.5.6}$$

as can readily be checked.

In manifestly twistorial terms the solution for the intersection point can be represented by the skew product of the two twistors. In particular, if we put

$$X^{\alpha\beta} = (Z_1^{\alpha}Z_2^{\beta} - Z_2^{\alpha}Z_1^{\beta})/(Z_1^{\alpha}Z_2^{\beta}I_{\alpha\beta}) \tag{2.5.7}$$

where a normalization factor has been included so as to ensure that $X^{\alpha\beta}I_{\alpha\beta} = 2$, then one finds

$$X^{\alpha\beta} = \begin{bmatrix} -\frac{1}{2}x_d x^d \varepsilon^{AB} & ix^{A}{}_{B'} \\ -ix_{A'}{}^{B} & \varepsilon_{A'B'} \end{bmatrix} \tag{2.5.8}$$

for the spinor parts of $X^{\alpha\beta}$, after a short calculation. Thus a spacetime point $x^{AA'}$ is represented in twistor terms by a simple skew-symmetric twistor $X^{\alpha\beta}$ (recall that "simple" here means that $X^{[\alpha\beta}X^{\gamma]\delta} = 0$) satisfying the normalization condition $X^{\alpha\beta}I_{\alpha\beta} = 2$, where $I_{\alpha\beta}$ is the infinity twistor defined in (2.3.6).

The dual description of the same spacetime point is formed by taking $X_{\alpha\beta} = \frac{1}{2}\varepsilon_{\alpha\beta\gamma\delta}X^{\gamma\delta}$. The complex conjugate spacetime point $\bar{x}^{a}$ is described dually by the complex conjugate twistor $\bar{X}_{\alpha\beta}$. The condition that a spacetime point should be real is that the dual twistor should be equal to the complex conjugate twistor, i.e. $X_{\alpha\beta} = \bar{X}_{\alpha\beta}$.

If $X^{\alpha\beta}$ and $Y^{\alpha\beta}$ represent, according to the description given above, the spacetime points x^a and y^a, respectively, then the quantity

$$-X^{\alpha\beta}Y_{\alpha\beta} = (x^a-y^a)(x_a-y_a) \tag{2.5.9}$$

is the norm of the spacetime separation of the two points. In particular, if $x^a = v^a-iw^a$ where v^a and w^a are real, then

$$\frac{1}{4} X^{\alpha\beta}\bar{X}_{\alpha\beta} = w_a w^a \tag{2.5.10}$$

is the norm of the imaginary part of x^a, which is a Poincaré invariant quantity.

That region of complex Minkowski space (CM) for which w^a is timelike and future-pointing will be called (notwithstanding some apparently unavoidable terminological awkwardness) the future - tube, and will be denoted CM^+. The region for which w^a is timelike and past-pointing will be denoted CM^-.

2.6 Projective Twistor Space.

An α-plane does not determine a twistor uniquely, but rather, as should be evident from equation (2.5.1), only up to an overall scale factor. An equivalence class of twistors all of which are proportional to each other is called a projective twistor, and by projective twistor space (PT) we mean the set of all such equivalence classes.

It is clear that projectively a twistor does not have a well-defined norm. Nevertheless, projectively the sign of the norm still makes sense, and thus PT can be divided into three parts denoted PT^+, PN, and PT^- according as to whether the norm $Z^\alpha\bar{Z}_\alpha$ is positive, zero, or negative.

Often we will use the twistor coordinates Z^α to denote the associated equivalence class in projective twistor space. In that case we refer to the components of Z^α as the homogeneous coordinates for the corresponding point in PT. The systematic use of homogeneous coordinates has the marvelous catalytic effect of simplifying much of the calculational work that crops up in algebraic geometry.

A point W_α in dual projective twistor space corresponds to a plane in PT. The plane consists of all those twistors Z^α that satisfy $W_\alpha Z^\alpha = 0$. Note that the equation for the plane is completely scale invariant.

The skew product $Z_1^{[\alpha}Z_2^{\beta]}$ between a pair of projective twistors Z_1^α and Z_2^α corresponds to the complex projective line (P^1) which joints them. Thus, lines in PT correspond to points in complex Minkowski space. If $\pi_1^{A'}$ is not proportional to $\pi_2^{A'}$, then we can normalize $Z_1^{[\alpha}Z_2^{\beta]}$ so as to obtain the convenient representation of space-time points given by equations (2.5.7) and (2.5.8). (If $\pi_1^{A'}$ is, in fact, proportional to $\pi_2^{A'}$, then the skew product $Z_1^{[\alpha}Z_2^{\beta]}$ represents a point at infinity.)

The representation of lines in PT (i.e. points in CM) by simple skew-symmetric twistors - these being the "Plücker coordinates" for the lines - allows us to derive a number of interesting results concerning the geometry of PT, several of which will be mentioned here:

2.6.1 Proposition. The intersection in PT of the line $X^{\alpha\beta}$ and the plane W_α is represented by the twistor $W_\alpha X^{\alpha\beta}$.

Proof. One must show that the twistor $W_\alpha X^{\alpha\beta}$ lies both on the line $X^{\alpha\beta}$ and the plane W_β. Clearly the latter holds, since $(W_\alpha X^{\alpha\beta})W_\beta = 0$. Now a twistor Z^β lies on the line $X^{\alpha\beta}$ if and only if $Z^{[\beta}X^{\gamma\delta]} = 0$. It follows therefore, from the quadratic p-relations $X^{\alpha[\beta}X^{\gamma\delta]} = 0$ (i.e. the simplicity conditions) by contraction with W_α, that $W_\alpha X^{\alpha\beta}$ lies on $X^{\alpha\beta}$. □

2.6.2 Proposition. The line $X^{\alpha\beta}$ lies entirely within PT^+ (that is to say, $Z^{[\alpha}X^{\beta\gamma]} = 0$ implies $Z^\alpha \bar{Z}_\alpha > 0$) if and only if the inequality

$$(2.6.3)\qquad (W_\alpha X^{\alpha\beta})(\bar{W}^\gamma \bar{X}_{\gamma\beta}) > 0$$

holds for every choice of a plane W_α.

Proof. By Proposition (2.6.1) above, $W_\alpha X^{\alpha\beta}$ represents the intersection of the plane W_α and the line $X^{\alpha\beta}$. Clearly, the line $X^{\alpha\beta}$ lies entirely within PT^+ if and only if its intersection with any plane lies in PT^+, which is precisely what (2.6.3) asserts. □

2.6.4 Proposition. Lines which lie entirely within PT^+ (the "top half" of projective twistor space) correspond to points in CM^+ (the future - tube).

Proof. The twistor Z^α lies on the line $X^{\alpha\beta}$ if and only if it is of the form $Z^\alpha = (ix^{AA'}\pi_{A'}, \pi_{A'})$ for some choice of $\pi_{A'}$. Writing $x^a = v^a - iw^a$, a short calculation establishes that

(2.6.5) $$Z^\alpha \bar{Z}_\alpha = 2w^{AA'}\bar{\pi}_A \pi_{A'} .$$

Now since $\bar{\pi}_A\pi_{A'}$ is null and future-pointing, it follows that $Z^\alpha\bar{Z}_\alpha > 0$ for all $\pi_{A'}$ if and only if w^a is timelike and future-pointing. □

Similarly, lines lying in $\overline{PT}^+ = PT^+ \cup PN$ correspond to points in the "closed" future tube $\overline{CM}^+$. Lines in $\overline{PT}^-$ correspond to points in $\overline{CM}^-$. Lines which intersect all three of PT^+, PN, and PT^- correspond to points for which w^a is spacelike. Real Minkowski space points correspond to lines in the hypersurface PN. For further discussion see Penrose 1967, section VI.

Chapter 2, Notes.

1. We require here various spinor identities, including the following:

$$-i\varepsilon_{abcd} = \varepsilon_{AC}\varepsilon_{BD}\varepsilon_{A'B'}\varepsilon_{C'D'} - \varepsilon_{AB}\varepsilon_{CD}\varepsilon_{A'C'}\varepsilon_{B'D'}$$

$$\omega^A = \varepsilon^{AB}\omega_B \; , \quad \omega_B = \omega^A\varepsilon_{AB} \; , \quad \pi^{A'} = \varepsilon^{A'B'}\pi_{B'} \; , \quad \pi_{B'} = \pi^{A'}\varepsilon_{A'B'}$$

$$\varepsilon^{AB} = -\varepsilon^{BA} \; , \quad \varepsilon^{[AB}\varepsilon^{C]D} = 0 \; , \quad \varepsilon^{A'B'} = -\varepsilon^{B'A'} .$$

See, for example, Pirani (1965), section 3, and Penrose (1968a) for standard expositions of spinor algebra. Proposition 2.1.2 appears in Penrose and MacCallum (1972), section 1.3.

2. For treatments of the global geometry of compactified Minkowski space, see Penrose (1963), Penrose (1965b) and, especially, Penrose (1965a). A somewhat more streamlined treatment is outlined in Penrose (1968a). Also see the account given in Hawking and Ellis (1973).

3. See Penrose (1967), section V.

CHAPTER 3

MASSIVE SYSTEMS AND THEIR INTERNAL SYMMETRIES

3.1 Momentum and Angular Momentum.

A massive system, like a massless system, is characterized by its total momentum and its total angular momentum. Unlike the case for a massless system, however, for a massive system it is not required that P^a and M^{ab} be related to one another directly in any special way. All that is required is that the momentum be timelike and future-pointing, and that the angular momentum behave appropriately under translations.

The angular momentum can be expressed in the form

(3.1.1) $$M^{ab} = \mu^{AB}\varepsilon^{A'B'} + \bar{\mu}^{A'B'}\varepsilon^{AB} \quad ,$$

where μ^{AB} is a symmetric spinor. Under a change of origin in complex Minkowski space the angular momentum is taken to transform as follows:

(3.1.2) $$\mu^{AB} \longrightarrow \mu^{AB} - P^{(A}{}_{B'}q^{B)B'} \quad ,$$

where $q^{AA'}$ is defined to be the position vector of the new origin with respect to the old. The <u>complex center of mass</u>[1] of the system is the set of all points in CM about which the angular momentum vanishes. It is, accordingly, given by those points $x^{AA'}$ which satisfy

(3.1.3) $$\mu^{AB} = P^{(A}{}_{B'}x^{B)B'} \quad .$$

The general solution to equation (3.1.3) is

(3.1.4) $$x^{AA'} = 2m^{-2}\mu^{AB}P_B{}^{A'} + \lambda P^{AA'} \quad ,$$

where m is the mass, and λ is an arbitrary complex number. The following result is illustrative of the significance of the complex center of mass:

<u>3.1.5 Proposition</u>. The spin-vector of a massive system is a measure of the system's displacement transverse to the momentum into the complex.

Proof. The spin-vector is defined, as usual, by $S^a = {*M}^{ab}P_b$. Writing ${*M}^{ab} = -i\mu^{AB}\varepsilon^{A'B'} + i\bar{\mu}^{A'B'}\varepsilon^{AB}$ one obtains

$$S^{AA'} = -i\mu^{AB}P_B{}^{A'} + i\bar{\mu}^{A'B'}P^A{}_{B'} \quad . \tag{3.1.6}$$

Combining (3.1.4) and (3.1.6) it follows directly that

$$x^a = v^a + i(\kappa P^a + m^{-2}S^a) \quad , \tag{3.1.7}$$

where $v^a = \frac{1}{2}(x^a + \bar{x}^a)$ and $\kappa = -i\frac{1}{2}(\lambda - \bar{\lambda})$. Thus $m^{-2}S^a$ is the transverse displacement of the center of mass into the complex. □

The complex center of mass will be discussed at greater length in Section 3.5, where an expression in explicit twistor terms will be derived for it.

3.2 The Kinematical Twistor.

The momentum and angular momentum of any system, massive or massless, can be expressed in terms of a certain type of symmetric twistor of valence $[{}^2_0]$ called the system's kinematical twistor. It is defined as follows:

$$A^{\alpha\beta} = \begin{pmatrix} -2i\mu^{AB} & P^A{}_{B'} \\ P_{A'}{}^B & 0 \end{pmatrix} \quad , \tag{3.2.1}$$

where $P^{AA'}$ is the momentum, and μ^{AB} is the angular momentum spinor. A necessary and sufficient condition for a symmetric twistor $A^{\alpha\beta}$ to be of the form (3.2.1) is that

$$A^{\alpha\beta}I_{\beta\gamma} = \bar{A}_{\beta\gamma}I^{\beta\alpha} \quad , \tag{3.2.2}$$

where $\bar{A}_{\alpha\beta}$ is the complex conjugate of $A^{\alpha\beta}$, given in the spinor parts by

$$\bar{A}_{\alpha\beta} = \begin{pmatrix} 0 & P_A{}^{B'} \\ P^{A'}{}_B & 2i\bar{\mu}^{A'B'} \end{pmatrix} \quad . \tag{3.2.3}$$

For a massive system we require that P^a be timelike and future pointing. This means that $A^{\alpha\beta}I_{\beta\gamma}W_\alpha\bar{W}^\gamma$ must be greater than zero for any choice of W_α. For a massless system, on the other hand, we require that $P_{AA'} = \bar{\pi}_A\pi_{A'}$ and $\mu_{AB} = i\omega_{(A}\bar{\pi}_{B)}$ for some choice of $Z^\alpha = (\omega^A , \pi_{A'})$. In fact, a kinematical twistor describes a massless system if and only if there exists a twistor Z^α such that

$$A^{\alpha\beta} = 2Z^{(\alpha}I^{\beta)\gamma}\bar{Z}_\gamma \quad , \tag{3.2.4}$$

which is equivalent to conditions (2.1.3) and (2.1.4).

3.3 The Decomposition of Massive Systems into Massless Subsystems.

A very curious fact about massive systems is that they can always be regarded as being composed out of a set of two or more massless subsystems. These massless subsystems are the "twistor constituents" of the associated massive system.

3.3.1 Theorem. For any integer $n > 1$, given a massive system with momentum P^a and angular momentum μ^{AB} one can find a set of n ZRM systems Z_i^α $(i = 1,\ldots,n)$ such that

$$P^a = \sum_i P_i^a \quad , \quad \mu^{AB} = \sum_i \mu_i^{AB} \quad , \tag{3.3.2}$$

where P_i^a and μ_i^{AB} label the momenta and angular momenta of the various ZRM systems described by Z_i^α.

Proof. First, it will be demonstrated how a massive system $\{P^a , \mu^{AB}\}$ can be decomposed into a pair of ZRM systems.

Let ξ^a be any unit spacelike vector $(\xi_a\xi^a = -1)$ orthogonal to P^a. Put:

$$P_1^a = (P^a + m\xi^a)/2 \quad , \quad P_2^a = (P^a - m\xi^a)/2 \quad . \tag{3.3.3}$$

It follows immediately that both of the momenta P_i^a (with $i = 1, 2$) are null, and thus that

$$P_1^a = \bar{\pi}^{1A}\pi_1^{A'} \quad , \quad P_2^a = \bar{\pi}^{2A}\pi_2^{A'} \tag{3.3.4}$$

for some choice of $\pi_1^{A'}$ and $\pi_2^{A'}$. (We have written $\bar{\pi}^{iA}$ for the complex conjugate of $\pi_i^{A'}$.) Now μ_{AB} , being symmetric, must be of the form $\alpha^{(A}\beta^{B)}$ for some α^A and β^B . Expanding β^B in the spinor basis generated by $\bar{\pi}^{1B}$ and $\bar{\pi}^{2B}$, one obtains

$$\beta^B = \theta_1\bar{\pi}^{1B} + \theta_2\bar{\pi}^{2B} \tag{3.3.5}$$

for some choice of θ_1 and θ_2. And then we have:

$$\mu^{AB} = \theta_1\alpha^{(A}\bar{\pi}^{B)1} + \theta_2\alpha^{(A}\bar{\pi}^{B)2} \quad . \tag{3.3.6}$$

Thus equations (3.3.2) hold, with P_i^a given as in (3.3.4), and with

$$\mu_1^{AB} = \theta_1\alpha^{(A}\bar{\pi}^{B)1} \quad , \quad \mu_2^{AB} = \theta_2\alpha^{(A}\bar{\pi}^{B)2} \quad . \tag{3.3.7}$$

Our two ZRM systems are accordingly then given by $Z_1^{\alpha} = (\theta_1\alpha^A , \pi_{1A'})$ and $Z_2^{\alpha} = (\theta_2\alpha^A , \pi_{2A'})$.

To decompose a massive system into <u>three</u> twistors, take a pair ξ^a , η^a of unit mutually orthogonal vectors lying in the 3-space orthogonal to the momentum and form the three null momenta P_i^a given by[(2)]:

$$\begin{aligned} P_1^a &= (P^a + \tfrac{m}{2}\xi^a + \tfrac{\sqrt{3}m}{2}\eta^a)/3 \\ P_2^a &= (P^a + \tfrac{m}{2}\xi^a - \tfrac{\sqrt{3}m}{2}\eta^a)/3 \\ P_3^a &= (P^2 - m\,\xi^a)/3 \quad , \end{aligned} \tag{3.3.8}$$

of which P^a is obviously the sum. Put $P_1^a = \bar{\pi}^{1A}\pi_1^{A'}$, $P_2^a = \bar{\pi}^{2A}\pi_2^{A'}$, and $P_3^a = \bar{\pi}^{3A}\pi_3^{A'}$ for an appropriate triplet of spinors $\pi_{A'i}$. Writing, as before, $\mu^{AB} = \alpha^{(A}\beta^{B)}$ one can clearly put $\beta^B = \sum_i \theta_i\bar{\pi}^{Bi}$ for some choice of θ_i. Hence, defining

$$\mu_1^{AB} = \theta_1\alpha^{(A}\bar{\pi}^{B)1} \text{ , etc.} \tag{3.3.9}$$

we see that equations (3.3.2) are satisfied, as desired. The three twistors Z_i^{α} are then defined by

$$Z_i^{\alpha} = (\theta_i\alpha^A , \pi_{iA'}) \quad . \tag{3.3.10}$$

This method can by iteration be extended to decompose a massive system into any number of twistors, as follows: Having split the massive system into three null subsystems, one now recombines two of the null subsystems to form a massive subsystem of the original system. Then that massive subsystem can be split into three null subsystems, giving us a splitting of the original system now into <u>four</u> null subsystems. This process can be repeated over and over until the desired number of null subsystems is achieved.□

Suppose that a massive system $A^{\alpha\beta}$ has been decomposed into a collection of massless subsystems denoted $A_i^{\alpha\beta}$ where i = 1, ..., n. On account of the linearity of the kinematical twistor in momentum and angular momentum it follows that $A^{\alpha\beta} = \sum_i A_i^{\alpha\beta}$. Each massless subsystem is described by a twistor Z_i^{α} for some value of the index i. The corresponding complex conjugate twistors will be denoted $\bar{Z}_{\alpha}^{i}$, raising the index i. It is useful to treat these indices according to the usual rules of tensor algebra, adopting the summation convention, and so forth. Then for the kinematical twistor of the complete massive system one has[3]:

$$A^{\alpha\beta} = 2Z_i^{(\alpha}I^{\beta)\gamma}\bar{Z}_{\gamma}^{i} \quad, \tag{3.3.11}$$

where now the contributions from all the various ZRM subsystems are automatically summed over.

3.4 Internal Symmetries.

Expression (3.3.11) can be regarded as the natural starting point for the development of the twistor approach to elementary particle physics. It shows how the momentum and angular momentum of a massive system can be built up out of a set of twistor constituents. According to Theorem (3.3.1) there will exist, for any kinematical twistor $A^{\alpha\beta}$, a set of twistor constituents Z_i^{α} such that $A^{\alpha\beta}$ is given by expression (3.3.11).

An important point to notice is that a massive system $A^{\alpha\beta}$ does not determine a <u>unique</u> set of twistor constituents. Linear transformations of the form

$$\begin{cases} Z_i^{\alpha} \longrightarrow R_{\beta i}^{\alpha j} Z_j^{\beta} + S_{ij}^{\alpha\beta}\bar{Z}_{\beta}^{j} \\ \bar{Z}_{\alpha}^{i} \longrightarrow \bar{R}_{\alpha j}^{\beta i}\bar{Z}_{\beta}^{j} + \bar{S}_{\alpha\beta}^{ij} Z_j^{\beta} \end{cases} \tag{3.4.1}$$

can be made such that—when $R_{\beta i}^{\alpha j}$ and $S_{ij}^{\alpha\beta}$ are suitably restricted—the kinematical twistor $A^{\alpha\beta}$, as given in equation (3.3.11), is left invariant.

<u>3.4.2 Theorem.</u> Linear transformations acting on $\pi_i^{A'}$ and $\bar{\pi}_A^{j}$ that preserve the momentum $\pi_i^{A'}\bar{\pi}_A^{i}$ are of the form:

$$\pi_i^{A'} \longrightarrow U_i^j \pi_j^{A'} \qquad \bar{\pi}^{iA} \longrightarrow \bar{U}_j^i \bar{\pi}^{jA} \quad , \tag{3.4.3}$$

where U_i^j is unitary (i.e. $U_i^j \bar{U}_j^k = \delta_j^k$).

<u>Proof</u>. It should be evident that if the momentum is to be preserved then $\pi_i^{A'}$, when transformed, must pick up no terms involving ω_i^A or $\bar{\omega}^{iA'}$. The most general linear transformation satisfying this condition, whilst maintaining the conjugacy relations between $\pi_i^{A'}$ and $\bar{\pi}^{iA}$ is given by

$$\left\{\begin{array}{l} \pi_i^{A'} \longrightarrow R_{iB'}^{A'j} \pi_j^{B'} + S_{ijB}^{A'} \bar{\pi}^{jB} \\ \bar{\pi}^{iA} \longrightarrow \bar{R}_{jB}^{Ai} \bar{\pi}^{jB} + \bar{S}_{B'}^{ijA} \pi_j^{B'} \end{array}\right. . \tag{3.4.4}$$

Under this transformation the momentum transforms as follows:

$$\begin{aligned} \pi_i^{A'} \bar{\pi}^{iA} \longrightarrow\ & (R_i^{A'j}{}_{B'} \bar{R}_{kC}^{Ai} + S_{ikC}^{A'} \bar{S}_{B'}^{ijA}) \pi_j^{B'} \bar{\pi}^{kC} \\ & + (R_i^{A'j}{}_{B'} \bar{S}_{C'}^{ikA}) \pi_j^{B'} \pi_k^{C'} + (\bar{R}_{jB}^{Ai} S_{ikC}^{A'}) \bar{\pi}^{jB} \bar{\pi}^{kC} \quad . \end{aligned} \tag{3.4.5}$$

In order for the momentum to be preserved, only the first of the three terms appearing on the right of (3.4.5) must survive, and the other two must vanish. Assuming that only the first term survives, then what is required is that

$$\pi_i^{A'} \bar{\pi}^{iA} = (R_i^{A'j}{}_{B'} \bar{R}_{kC}^{Ai} + S_{ikC}^{A'} \bar{S}_{B'}^{ijA}) \pi_j^{B'} \bar{\pi}^{kC} \tag{3.4.6}$$

for all values of $\pi_i^{A'}$. This must, in particular, be true when $\pi_i^{A'}$ is of the following degenerate form, suggested by the "Segre embedding" (cf. Mumford 1976, section 2B):

$$\pi_i^{A'} = \pi^{A'} q_i \quad , \tag{3.4.7}$$

for some value of $\pi^{A'}$ and q_i. Substituting (3.4.7) into (3.4.6) we then obtain

$$q_i \bar{q}^i \pi^{A'} \bar{\pi}^{A} = R_i^{A'} \bar{R}^{Ai} + S_i^{A'} \bar{S}^{Ai} \quad , \tag{3.4.8}$$

where $R_i^{A'}$ and $S_i^{A'}$ are defined by

$$R_i^{A'} = R_i^{A'j}{}_{B'} \pi^{B'} q_j \quad , \quad \bar{R}^{Ai} = \bar{R}_{jB}^{Ai} \bar{\pi}^{B} \bar{q}^j \quad , \tag{3.4.9}$$

(3.4.10) $$S_i^{A'} = S_{ijB}^{A'}\bar{\pi}^B\bar{q}^j \quad , \quad \bar{S}^{Ai} = \bar{S}_{B'}^{ijA}\pi^{B'}q_j \quad .$$

Now note that (3.4.8) is, on the left-hand side, a future-pointing null vector, whereas on the right-hand side is a <u>sum</u> of future-pointing null vectors. It follows that <u>each</u> of these null vectors must be proportional (with a positive or zero factor of proportionality) to $\pi^{A'}\bar{\pi}^A$. This will be the case only if we have

(3.4.11) $$R_i^{A'} = \pi^{A'}r_i \quad , \quad \bar{R}^{Ai} = \bar{\pi}^A\bar{r}^i \quad ,$$

(3.4.12) $$S_i^{A'} = \pi^{A'}s_i \quad , \quad \bar{S}^{Ai} = \bar{\pi}^A\bar{s}^i \quad ,$$

for all values of $\pi^{A'}$, with appropriate choices of r_i and s_i. Equation (3.4.12), however, is incompatible with equation (3.4.10) unless s_i vanishes; thus $S_{ijB}^{A'}$ vanishes—since $S_i^{A'}$ will vanish for all values of $\bar{\pi}^A$ and $\bar{q}^k$ only if $S_{ijB}^{A'}$ itself is zero. One can infer now, moreover, that the last two terms in (3.4.5) are zero, as desired. Equations (3.4.8), (3.4.9), and (3.4.11) taken in conjunction then give

(3.4.13) $$R_{i\ B'}^{A'j} = \varepsilon_{B'}{}^{A'}U_i^j \quad ,$$

with U_i^j unitary. □

<u>3.4.14 Theorem.</u> Linear transformations acting on Z_i^α and $\bar{Z}_\alpha^i$ that preserve the momentum $\pi_i^{A'}\bar{\pi}^{iA}$ and the angular momentum $i\omega_j^{(A}\bar{\pi}^{B)j}$ and also preserve the conjugacy relations between Z_i^α and $\bar{Z}_\alpha^i$, are of the form

(3.4.15) $$\pi_i^{A'} \longrightarrow U_i^j\pi_j^{A'} \quad ,$$

(3.4.16) $$\omega_i^A \longrightarrow U_i^j(\omega_j^A + \Lambda_{jk}\bar{\pi}^{kA}) \quad ,$$

together with the corresponding complex conjugate transformations, U_i^j being an arbitrary unitary matrix, and Λ_{jk} being an arbitrary skew-symmetric matrix. In manifest twistor terms these transformations are given by

(3.4.17) $$Z_i^\alpha \longrightarrow U_i^j(Z_j^\alpha + I^{\alpha\beta}\Lambda_{jk}\bar{Z}_\beta^k) \quad ,$$

together with its complex conjugate; the set of all such transformations forms a group.

Proof[4]. From Theorem 3.4.2 we know already that (3.4.15) is the most general transformation acting on $\pi_i^{A'}$ that preserves the momentum. It should be evident that a transformation acting on ω_i^A that preserves $\omega_i^{(A}\bar{\pi}^{B)i}$ must only contain terms linear in ω_i^A and $\bar{\pi}^{Bi}$, and must contain no terms involving $\bar{\omega}^i_{A'}$, or $\pi_i^{A'}$. Therefore, transformations of the form

$$\omega_i^A \longrightarrow R_{iB}^{Aj}\,\omega_j^B + S_{ijB}^A\,\bar{\pi}^{jB} \qquad (3.4.18)$$

are considered. Transformation (3.4.18), taken in conjunction with $\bar{\pi}^{Bi} \longrightarrow \bar{U}^i_j\bar{\pi}^{Bj}$, gives

$$\omega_i^{(A}\bar{\pi}^{B)i} \longrightarrow \bar{U}^i_k R_i^j{}^{(A}_C\,\bar{\pi}^{B)k}\omega_j^C + \bar{\pi}^{k(A}S^{B)}_{ijC}\bar{U}^i_k\bar{\pi}^{jC} \qquad (3.4.19)$$

Thus, for the preservation of the angular momentum we require that R_{iB}^{Aj} and S_{ij}^{AB} satisfy the following relations:

$$\omega_i^{(A}\bar{\pi}^{B)i} = \bar{U}^i_k R_i^j{}^{(A}_C\,\bar{\pi}^{B)k}\omega_j^C \quad , \qquad (3.4.20)$$

$$\bar{\pi}^{k(A}S^{B)}_{i\,jC}\bar{U}^i_k\bar{\pi}^{jC} = 0 \quad , \qquad (3.4.21)$$

for all values of ω_i^A and $\bar{\pi}^{iA}$. In particular, putting $\pi_i^{A'} = \pi^{A'}q_i$, these requirements become:

$$\bar{q}^i\omega_i^{(A}\bar{\pi}^{B)} = \bar{q}^k\bar{U}^i_k R^j_{iC}{}^{(A}\bar{\pi}^{B)}\omega_j^C \quad , \qquad (3.4.22)$$

$$\bar{q}^j\bar{q}^k\bar{\pi}^{(A}S^{B)}_{ijC}\bar{U}^i_k\bar{\pi}^C = 0 \quad . \qquad (3.4.23)$$

Equation (3.4.22) with no further ado implies

$$R_{iB}^{jA} = U_i^j\varepsilon_B{}^A \quad . \qquad (3.4.24)$$

Equation (3.4.23) implies, with a bit of algebra,

$$S^{AB}_{i(j}\bar{U}^i_{k)} = 0 \quad , \qquad (3.4.25)$$

which, when substituted back into (3.4.21), gives, after a short calculation:

$$S^A_{ijB} = U^k_i\Lambda_{kj}\varepsilon_B{}^A \quad , \qquad (3.4.26)$$

with Λ_{ij} skew-symmetric. Substituting (3.4.26) and (3.4.24) into (3.4.18), we obtain (3.4.16), as desired. The proof that (3.4.15) and (3.4.16) are equivalent to (3.4.17) is quite routine, and can be left to the reader. To prove that transformations of the form (3.4.17) form a group is straightforward. One must show that if $g_1 = (U_1 , \Lambda_1)$ is one such transformations, and $g_2 = (U_2 , \Lambda_2)$ is another, then their composition $g_2 g_1$ is yet another element $g_3 = (U_3 , \Lambda_3)$. Indeed, one finds $(U_3 , \Lambda_3) = (U_2 U_1 , \Lambda_1 + \bar{U}_1 \Lambda_2 \bar{U}_1)$. Moreover, one finds $g = (U , \Lambda)$ always has a unique inverse—namely: $g^{-1} = (\bar{U}, -U\Lambda U)$. □

Those transformations for which $\Lambda = 0$ form a subgroup, referred to as the "unitary" part of the internal symmetry group. Transformations for which $\Lambda \neq 0$ are called "inhomogeneous transformations". We shall refer to the subgroup of transformations for which $U_i^j = \delta_i^j$ as the group of "restricted" inhomogeneous transformations. It is a tenet of the so-called twistor particle hypothesis that these n-twistor internal symmetry groups are closely related to the various phenomenological symmetries that arise in the course of the investigation of elementary particles and their interactions (cf. Penrose, 1977).

3.5 The Center of Mass Twistor.

In connection with the description of a massive system $A^{\alpha\beta}$ in terms of a set of twistor constituents Z_i^α it is useful to introduce the quantities

(3.5.1) $$M_{ij} = Z_i^\alpha Z_j^\beta I_{\alpha\beta} \quad , \quad \bar{M}^{ij} = \bar{Z}_\alpha^i \bar{Z}_\beta^j I^{\alpha\beta} \; ,$$

called partial mass tensors. The name derives from the validity of the formula

(3.5.2) $$m^2 = M_{ij}\bar{M}^{ij} \quad ,$$

which expresses the squared mass of the system in terms of these tensors.

Now consider the skew-symmetric twistor $R^{\alpha\beta}$ defined by

(3.5.3) $$R^{\alpha\beta} = 2m^{-2} Z_j^\alpha Z_k^\beta \bar{M}^{jk} \quad .$$

Using (3.5.1) and (3.5.2) it is straightforward to verify that $R^{\alpha\beta}$ satisfies the normalization condition

$$R^{\alpha\beta}I_{\alpha\beta} = 2 \quad . \tag{3.5.4}$$

Moreover, we can easily see that $R^{\alpha\beta}$ is simple, for equation (3.5.3) can be rewritten in the form

$$R^{\alpha\beta} = 2m^{-2}Y^{\alpha}_{A}Y^{\beta A} \quad , \tag{3.5.5}$$

where Y^{α}_{A} is defined by:

$$Y^{\alpha}_{A} = Z^{\alpha}_{j}\bar{\pi}^{\bar{j}}_{A} \quad . \tag{3.5.6}$$

Equation (3.5.5) confirms that $R^{\alpha\beta}$ is indeed the skew product of a pair of twistors of valence one.

Since $R^{\alpha\beta}$ is simple and satisfies equation (3.5.4), it represents—according to the description elaborated in Section 2.5—a point in complex Minkowski space. In fact, we have the following result:

<u>3.5.7 Theorem</u>. The twistor $R^{\alpha\beta}$ represents a point on the system's complex center of mass. Moreover, by subjecting Z^{α}_{i} to inhomogeneous transformations, the point corresponding to $R^{\alpha\beta}$ can be translated so as to range over the entirety of the system's center of mass.

<u>Proof</u>. We shall write (cf. equation 2.5.8):

$$R^{\alpha\beta} = \begin{bmatrix} -\frac{1}{2}R_{d}R^{d}\varepsilon^{AB} & iR^{A}{}_{B'} \\ -iR_{A'}{}^{B} & \varepsilon_{A'B'} \end{bmatrix} \tag{3.5.8}$$

for the spinor parts of $R^{\alpha\beta}$. Then using definition (3.5.3) we have:

$$\begin{aligned} iR^{A}{}_{B'} &= 2m^{-2}\omega^{A}_{i}\pi_{jB'}\bar{M}^{\bar{i}\bar{j}} \\ &= 2m^{-2}\omega^{A}_{i}\pi_{jB'}\bar{\pi}^{\bar{i}}_{B}\bar{\pi}^{\bar{j}B} \\ &= -2m^{-2}\omega^{(A}_{i}\pi^{B)\bar{i}\bar{j}}\pi_{B}\pi_{jB'} -2m^{-2}\omega^{[A}_{i}\pi^{B]\bar{i}\bar{j}}\pi_{B}\pi_{jB'} \\ &= 2im^{-2}\mu^{AB}P_{BB'} - i\lambda_{0}P^{A}_{B'} \quad , \end{aligned} \tag{3.5.9}$$

where $\mu^{AB} = i\omega_j^{(A}\pi^{B)j}$, and λ_0 is given by the formula:

(3.5.10) $$\lambda_0 = -im^{-2}\omega_i^{A}\pi_A^{i} \quad .$$

Comparing (3.5.9) with (3.1.4) we see that $R^{AA'}$ is indeed on the complex center of mass of the Z_i^{α} system.

Inspection of (3.5.3) makes it evident that $R^{\alpha\beta}$ is invariant under unitary transformations acting on Z_i^{α}. It is not, however, invariant under restricted inhomogenous transformations. In fact, it may not be entirely obvious that $R^{\alpha\beta}$ remains simple when it is subjected to a transformation of the form

(3.5.11) $$Z_i^{\alpha} \longrightarrow Z_i^{\alpha} + I^{\alpha\beta}\Lambda_{ij}\bar{Z}^j \quad .$$

To verify that $R^{\alpha\beta}$ does indeed remain simple under (3.5.11), note that Y_B^{α} , defined in equation (3.5.6), undergoes the transformation

(3.5.12) $$Y_B^{\alpha} \longrightarrow \tilde{Y}_B^{\alpha} \quad , \quad \tilde{Y}_B^{\alpha} := Y_B^{\alpha} - \frac{1}{2} I^{\alpha}{}_B \bar{M}^{ij}\Lambda_{ij} \quad ,$$

where $I^{\alpha}{}_B = (\varepsilon^A{}_B \ , 0)$. And thus $R^{\alpha\beta}$ transforms according to the formula

(3.5.13) $$R^{\alpha\beta} \longrightarrow 2m^{-2}\tilde{Y}_A^{\alpha}\tilde{Y}^{\beta A} \quad ,$$

from which it is clear that $R^{\alpha\beta}$ remains simple.

Having verified that $R^{\alpha\beta}$ remains a point, it suffices to examine the expression for $R^{AA'}$ now, in order to see how $R^{AA'}$ behaves under an inhomogeneous transformation. The last line of (3.5.9) shows us that λ_0 is the only quantity apt to alter under the effects of (3.5.11). And using (3.5.10) we deduce that λ_0 transforms as follows:

(3.5.14) $$\lambda_0 \longrightarrow \lambda_0 + \lambda \quad , \quad \lambda := im^{-2}\bar{M}^{ij}\Lambda_{ij} \quad .$$

By an appropriate choice of Λ_{ij} (e.g. $\Lambda_{ij} = -i\lambda M_{ij}$) one can then translate $R^{\alpha\beta}$ to any point on the complex center of mass of the system. □

If all n of a set of twistors Z_i^{α} happen to pass through the same spacetime point $x^{AA'}$, then that spacetime point, as one might justifiably suspect, lies on the center of mass of the system. In order to formalize this observation it is useful to introduce an operator ρ_x defined by

(3.5.15) $$\rho_x Z^\alpha = -X^{\alpha\beta} I_{\beta\gamma} Z^\gamma \quad ,$$

where $X^{\alpha\beta}$ corresponds—according to equation (2.5.8)—to the spacetime point $x^{AA'}$. We call ρ_x the "restriction operator" for the spacetime point $x^{AA'}$. Note that equation (3.5.15) can alternatively be written in the form

(3.5.16) $$\rho_x Z^\alpha = X^{\alpha B'} \pi_{B'} \quad , \quad X^{\alpha B'} = (ix^{AB'} \, , \, \varepsilon_{A'}{}^{B'}) \quad ,$$

and also that we have

(3.5.17) $$\rho_x Z^\alpha = (ix^{AA'} \pi_{A'} \, , \, \pi_{A'}) \quad ,$$

showing that $x^{AA'}$ does indeed lie in the α-plane determined by the twistor $\rho_x Z^\alpha$.

<u>3.5.18 Proposition</u>. The center of mass twistor $R^{\alpha\beta}$ satisfies the relation:

(3.5.19) $$\rho_x R^{\alpha\beta} = X^{\alpha\beta} \quad .$$

<u>Proof</u>. This result should, in fact, be clear on geometrical grounds. It can be directly verified as follows. Let $X^{\alpha B'}$ be defined as in (3.5.16). It is straightforward to deduce, using (2.5.8), that

(3.5.20) $$X^{\alpha\beta} = X^\alpha{}_{B'} X^{\beta B'} \quad .$$

Then we have:

(3.5.21)
$$\begin{aligned}
\rho_x R^{\alpha\beta} &= 2m^{-2} \rho_x Z_j^\alpha Z_k^\beta \bar{M}^{jk} \\
&= 2m^{-2} X^{\alpha A'} X^{\beta B'} \pi_{A'j} \pi_{B'k} \bar{M}^{jk} \\
&= m^{-2} X^{\alpha A'} X^{\beta B'} \varepsilon_{A'B'} M_{jk} \bar{M}^{jk} \\
&= X^{\alpha\beta} \quad ,
\end{aligned}$$

where we have used (3.5.16) and (3.5.2). □

Chapter 3, Notes.

1. See Newman and Winicour (1974), Tod and Perjés (1977), Tod (1975), and Tod (1977).

2. Cf. Feynman, Kislinger, and Ravndal (1971), particularly their formula 4a.

3. See Penrose and MacCallum (1972), p. 308—where formula (3.3.11) first appears—for further discussion.

4. This result has a long and interesting history, to which many individuals—Penrose, Perjés, Sparling, Hodges, and Tod, to name a few—have contributed. The twistor internal symmetry groups were being discussed extensively as early as the Spring of 1973—although they were not being called "internal symmetry groups" yet, at that time—in seminars at Birkbeck College, London. Theorem 3.4.14 was conjectured during that period—and believed by most of us to be valid—although rigorous justification was not forthcoming until 1977 by Penrose and Sparling (for the case of infinitesimal transformations) and 1978 by Sparling (for the general case). Early references to the twistor internal symmetry groups include Penrose (1975a), pp. 328-329; Penrose (1975b); and Perjés (1975).

CHAPTER 4

TWISTOR QUANTIZATION: ZERO REST MASS FIELDS

4.1 What is Twistor Quantization?

Populated as it is with all its curious symbols, enigmatic diagrams, and bizarre hieroglyphs, it is no wonder that the literature of twistor theory continues to confound and baffle the outside world—and it is no wonder that the mysterious and cultish procedure of "twistor quantization" remains, in the eyes of the masses, a kind of cryptic rite, to be performed in the dead of night, surrounded by the exotic paraphenalia of algebraic geometry, sheaf cohomology, and intuitionistic logic.

More seriously, what we mean by twistor quantization is a set of rules according to which the variables of classical twistor geometry are systematically replaced by certain operators, out of which one constructs quantum mechanical observables acting on an associated Hilbert space, the elements of which are interpreted as particle states.

For the sake of clarity, it should be emphasized that twistor quantization is, at least as presently formulated, a "first quantization" technique: that is to say, it allows for the passage from an essentially classical and finite dimensional picture to the elementary quantum mechanics of single particle systems. In twistor terms particle states are represented in terms of classes of holomorphic functions defined over certain domains of twistor space and products of twistor spaces. Physical observables are represented by holomorphic differential operators acting on these functions. It is our hypothesis that all elementary particle states can be described in this way: thus, through the examination of the complex analytic geometry of twistor space a classification scheme for particles emerges.

4.2 The Helicity Operator.

Within the purely classical aspect of twistor theory, as outlined in the previous two chapters, twistors and their complex conjugates are treated on more or less an equal footing. The passage to the quantum theory is achieved with the elimination of the $\bar{Z}_\alpha$ variables: wherever the complex conjugate variable $\bar{Z}_\alpha$ occurs, it is replaced by the holomorphic differential operator $\hat{Z}_\alpha = -\partial/\partial Z^\alpha$. It should be stressed that there is no single definitive reason for introducing the substitution $\bar{Z}_\alpha \longrightarrow \hat{Z}_\alpha$. There are, however, many different indications from many different viewpoints that lead, ultimately, to the conclusion that this is the correct way to proceed for quantization.

As described in Section 2.1, the classical expression for the helicty of a ZRM system is

(4.2.1) $$s = \frac{1}{4}(Z^\alpha\bar{Z}_\alpha + \bar{Z}_\alpha Z^\alpha) \quad .$$

Note that the expression for the helicity has been symmetrized in its dependence on Z^α and $\bar{Z}_\alpha$, as is usually done—at this stage of the analsyis no "factor ordering" problems ever arise—in the preparation of a classical dynamical variable for quantization. When the quantization rule

(4.2.2) $$\bar{Z}_\alpha \longrightarrow \hat{Z}_\alpha \ (= -\partial/\partial Z^\alpha)$$

is brought into effect, we find that the commutation relations

(4.2.3) $$[Z^\alpha , \hat{Z}_\beta] = \delta^\alpha_\beta$$

imply that the <u>helicity operator</u> $\hat{s}$ should be given by

(4.2.4) $$\hat{s} = \frac{1}{2} Z^\alpha\hat{Z}_\alpha - 1 \quad ,$$

or, equivalently, by

(4.2.5) $$-2\hat{s}-2 = Z^\alpha\partial/\partial Z^\alpha \quad .$$

One minor question. We have an operator: but what on earth does it act on? When we started off s was a classical dynamical observable for a classical system of zero rest mass. Upon quantization we obtain an operator $\hat{s}$ which is to be interpreted as a quantum mechanical observable acting on zero rest mass quantum mechanical systems; thus, the functions upon which $\hat{s}$ operates are to be interpreted as zero rest mass particle states.

The helicity operator depends only upon the variable Z^α. It is consistent with this fact and with the general principles of twistor quantization to choose the states on which $\hat{s}$ acts to be holomorphic functions, defined over some domain of twistor space. (We shall return shortly to the question of what sort of domain should, in fact, be taken—this being, in reality, a rather involved and intricate question.) Now in equation (4.2.5) one observes the appearance of the Euler homogeneity operator $Z^\alpha\partial/\partial Z^\alpha$. Thus, according to formula (4.2.5), in order to describe a ZRM field of helicity s (where s now denotes the eigenvalue of the operator $\hat{s}$) we should use a holomorphic function $f(Z^\alpha)$, homogeneous of degree -2s-2 in Z^α, i.e. satisfying

$$f(\lambda Z^\alpha) = \lambda^{-2s-2} f(Z^\alpha) \quad . \tag{4.2.6}$$

And indeed it emerges that in terms of such twistor functions one can give a complete and compelling description of massless fields—at least insofar as we confine out interests to analytic free fields defined on suitable domains of complex Minkowski space.

4.3 Positive Helicity Fields.

By a zero rest mass free field one means a solution to one of the following three equations(1):

$$\nabla^{AA'}\phi_{A'B'\ldots C'} = 0 \quad , \tag{4.3.1}$$

$$\Box\phi = 0 \quad , \tag{4.3.2}$$

$$\nabla^{A'A}\phi_{AB\ldots C} = 0 \quad . \tag{4.3.3}$$

We are primarily interested in positive frequency fields that furthermore exhibit the property of being analytic throughout the closure of the forward tube of complex Minkowski space—the "closed" forward tube being defined by: $x^a \in \overline{CM}^+$ iff $x^a = v^a - iw^a$ with w^a timelike future-pointing, null future-pointing, or zero(2). In the event that positive frequency fields are under consideration equations (4.3.1), (4.3.2), and (4.3.3) describe, respectively, fields of positive, zero, and negative helicity—for negative frequency fields (i.e., fields analytic throughout CM^-) the helicities are reversed. In each case, the spin is equal to one-half the number of spinor indices appearing on the field (the fields being assumed to be symmetric in all their indices).

Now we consider the problem of finding solutions of (4.3.1). In this connection it is useful to recall the notation introduced towards the end of Section 3.5, expressed in equation (3.5.17). For the restriction of a holomorphic function to the region of twistor space corresponding to a spacetime point $x^{AA'}$ we write

$$\rho_x f(Z^\alpha) = f(\rho_x Z^\alpha) = f(ix^{AA'}\pi_{A'}, \pi_{A'}) \quad . \tag{4.3.4}$$

Thus $\rho_x f(Z^\alpha)$ is to be regarded as a function of $x^{AA'}$ and $\pi_{A'}$. Equation (4.3.1) can be solved by means of the following contour integral formula:

$$\phi_{A'B'\ldots C'} = \oint \rho_x \pi_{A'}\pi_{B'}\ldots\pi_{C'} f(Z^\alpha)\Delta\pi \quad , \tag{4.3.5}$$

where the differential form $\Delta\pi$ is defined by

$$\Delta\pi = \pi_{A'}d\pi^{A'} \quad . \tag{4.3.6}$$

The twistor function is taken—in accordance with the discussion of the helicity operator in the previous section—to be homogeneous of degree $-2s-2$, where the number of spinor indices appearing on $\phi_{A'B'\ldots C'}$ is $2s$. Since $\Delta\pi$ is homogeneous of degree 2 in $\pi_{A'}$ it follows that the expression under the integral sign in (4.3.5) is—in its entirety—homogeneous of degree zero. The idea now is that the π-dependence is integrated away, leaving behind a field that depends upon $x^{AA'}$ above.

4.3.7 Proposition[3]. The field $\phi_{A'B'\ldots C'}$ defined in (4.3.5) automatically satisfies the ZRM equation (4.3.1).

Proof. First we require the following formula, based on the differential chain rule:

$$\begin{aligned} \nabla_{AA'}\rho_x f(Z^\alpha) &= \nabla_{AA'} f(ix^{BB'}\pi_{B'}, \pi_{B'}) \\ &= -i(\nabla_{AA'}x^{BB'}\pi_{B'})\rho_x \hat{\pi}_B f(Z^\alpha) \\ &= -i\varepsilon_A{}^B\varepsilon_{A'}{}^{B'}\pi_{B'}\rho_x\hat{\pi}_B f(Z^\alpha) \end{aligned} \tag{4.3.8}$$

$$= -i\rho_x \pi_{A'} \hat{\pi}_A f(Z^\alpha) \quad ,$$

where, compatible with the identification $\hat{Z}_\alpha = -\partial/\partial Z^\alpha$, we have put $\hat{\pi}_A = -\partial/\partial\omega^A$. Differentiating (4.3.5) and using (4.3.8), one finds that (4.3.1) follows immediately (on account of the trivial identity $\pi^{A'}\pi_{A'} = 0$). □

It should not come as a surprise that, when acting on twistor functions, the following identity is valid:

(4.3.9) $$i\nabla_{AA'}\rho_x = \rho_x \hat{P}_{AA'} \quad ,$$

where $\hat{P}_{AA'} = \hat{\pi}_A \pi_{A'}$ is the momentum operator, obtainable from formula (2.1.3) by the application of twistor quantization, i.e. by means of the substitution

(4.3.10) $$\bar{\pi}_A \longrightarrow \hat{\pi}_A \ (= -\partial/\partial\omega^A) \quad .$$

Indeed, if the momentum and angular momentum appearing in equations (2.1.3) and (2.1.4) are quantized by means of (4.3.10) together with

(4.3.11) $$\bar{\omega}^{A'} \longrightarrow \hat{\omega}^{A'} \ (= -\partial/\partial\pi_{A'}) \quad ,$$

then the resulting operators $\hat{P}^a$ and $\hat{M}^{ab}$ satisfy the commutation relations

(4.3.12) $$[\hat{P}^a \, , \, \hat{P}^b] = 0 \quad , \quad [\hat{P}^a \, , \, \hat{M}^{bc}] = -2ig^{a[b}\hat{P}^{c]} \quad ,$$
$$[\hat{M}^{ab} \, , \, \hat{M}^{cd}] = 2i\hat{M}^{c[a}g^{b]d} - 2i\hat{M}^{d[a}g^{b]c} \quad ,$$

which are the correct quantum mechanical commutation relations for momentum and angular momentum operators—i.e. they exhibit the correct commutation relations for the generators of the Lie algebra of the Poincaré group.

Summing up: A twistor function $f(Z^\alpha)$ represents—with some "gauge freedom" (i.e. cohomological coboundary freedom) corresponding to adding onto $f(Z^\alpha)$ any function which integrates to zero with respect to the contour chosen in equation (4.3.5), irrelevant as far as the present discussion is concerned, although ultimately of considerable significance—the quantum mechanical wave function of a zero rest mass particle; the state—assuming positive frequency (cf. Section 4.5)—has helicity s if the twistor function is homogeneous of degree -2s-2 (the negative helicity case will be treated in the next section); a contour integral formula can be employed to relate the twistor function description of the state to the more conventional spacetime field description.

4.4 Negative Helicity Fields.

For a number of years various aspects of twistor theory were hampered in their development on account of the fact that although twistor functions $f(Z^\alpha)$ could be used to generate positive frequency positive helicity solutions of the ZRM equations, it was not known how to use such functions in order to generate positive frequency negative helicity solutions: in Penrose and MacCallum (1972), for example, one finds on p. 261 the remark : "... it turns out that spinor fields with unprimed indices, whose positive energy parts represent left-handed particles, must be represented by functions on the dual twistor space, while spinors with primed indices, whose positive energy parts represent right-handed particles will be represented by functions on the twistor space." Thus it seemed that positive helicity fields were to be described by functions $f(Z^\alpha)$ with homogeneity lesser than or equal to -2, while negative helicity fields were to be described by functions $g(W_\alpha)$, again with homogeneity lesser than or equal to -2, on dual twistor space. The twistor functions with homogeneity greater than -2 were apparently to be ignored and discarded—in fact, they were instead relegated to a rather curious and mysterious status, and were called "active" twistor functions (as opposed to the other ones, which were called "passive") on account of the prominent role they played in the description of certain types of scattering processes [cf. Penrose and MacCallum (1972), p. 277]. This state of affairs, despite the fact that it generated an enormous amount of good conversation and interesting philosophy, was unsatisfactory in some respects.

However, the matter was resolved in 1973 when a "direct" method (i.e., a contour integral formula) was formulated for extracting the field information of "active" twistor functions. The correct expression is given by

$$\text{(4.4.1)} \qquad \phi_{AB\ldots C} = \oint \rho_x \hat{\pi}_A \hat{\pi}_B \ldots \hat{\pi}_C f(Z^\alpha) \Delta\pi \quad ,$$

where $f(Z^\alpha)$ is homogeneous of degree -2s-2, with s negative, the operator

$\hat{\pi}_A$ $(= -\partial/\partial\omega^A)$ appearing $-2s$ times in the integrand.

4.4.2 Proposition. The field $\phi_{AB\ldots C}$ defined in (4.4.1) automatically satisfies the ZRM equation (4.3.3).

Proof. Applying formula (4.3.8), equation (4.3.3) immediately follows on account of the trivial operator identity $\hat{\pi}_A\hat{\pi}^A = 0$. □

In spite of the fact that (4.4.1) evidently gave a correct description of negative helicity fields that when taken in conjunction with (4.3.5) allowed for a complete characterization of massless fields of all helicities in terms of twistors functions all defined on the same space (i.e. the Z^α space), the "old" description (i.e. the Z^α space for one helicity, and the W_α space for the other) was adhered to for a long time : see, for example, Woodhouse (1975) and the well-known opus Penrose (1975a)—it was only with the coming of the "twistor particle hypothesis" [Penrose 1976], and the "twisted photon" [Ward 1977a] that the validity and appropriateness of formula (4.4.1) for negative helicity massless fields emerged as more clearly obvious: complete justification followed later with the advent of sheaf cohomology methods. (Cf. Chapter 10.)

4.5 The Positive Frequency Condition.

Twistor contour integral formulae for massless fields—alluded to briefly on page 347 of Penrose's 1967 "Twistor Algebra" paper—are first introduced and described in some detail in Penrose (1968b), in which it is asserted, on page 64, that a "... feature of twistor analysis, which has been highly instrumental in the motivation for its original development, lies in the extent to which it 'geometrises' an important aspect of quantum mechanics, namely the splitting of field amplitudes into positive and negative frequency parts." And today, notwithstanding numerous advances and developments, this feature of the theory remains one of its most fascinating and compelling—especially when taken, according to various proposals, over to the non-linear regime [cf. Penrose 1976; and Ward 1977a and 1977b].

A plane wave $e^{-ik\cdot x}$ has positive energy if k is future-pointing, and negative energy if k is past-pointing. In the positive energy case the wave can be extended over all of $\overline{CM}^+$ since, putting $x^a = v^a - iw^a$, the wave function in $\overline{CM}^+$ is given by $e^{-ik\cdot v}e^{-k\cdot w}$, and is manifestly well-behaved for all values of future-pointing w. Similarly, in the negative energy case the wave can be extended over all of $\overline{CM}^-$. It is common, in many branches of theoretical physics, to define positive and negative frequency in terms of these analyticity conditions——and this is the case for twistor theory as well.

According to Proposition 2.6.4, and the remarks immediately following it, the points of $\overline{CM}^+$ correspond to projective lines lying in $\overline{PT}^+$, i.e. the domain for which $Z^\alpha\bar{Z}_\alpha \geq 0$. Thus in order to characterize a massless field as positive frequency it suffices to restrict ones attention to $\overline{PT}^+$. By requiring $f(Z^\alpha)$ to have suitable analyticity properties in $\overline{PT}^+$ one can ensure that the related field is well-defined throughout $\overline{CM}^+$, and thus of positive frequency.

This brings us back to the problem, mentioned in Section 4.3, of the domain on which $f(Z^\alpha)$ is to be defined, and the question of what sort of singularities $f(Z^\alpha)$ ought to exhibit. Part of the reason why this problem is so difficult is due to the fact that the question is not posed quite correctly. For the sort of objects that are being dealt with here are not really functions at all—at least, in the standard sense—but rather, are elements of the sheaf cohomology group

$$H^1(\overline{PT}^+, O(-2s-2)) \quad , \tag{4.5.1}$$

where $O(-2s-2)$ is the sheaf of germs of holomorphic functions, homogeneous of degree $-2s-2$. A twistor function $f(Z^\alpha)$ provides a "representative cocycle" for an element of (4.5.1). It is possible to have several distinct twistor functions, all defined over distinct domains, all of which yet are representatives for the same element of the cohomology group (4.5.1)—this is why, in the older twistor literature, for a specified ZRM field the domain of the corresponding twistor function seems a bit "shifty". And it was only in 1976 that the matter began to clear up, and it emerged that positive frequency analytic massless fields of helicity s corresponded to elements of (4.5.1), thereby specifying precisely the relationship between such fields and the complex analytic geometry of twistor space. For discussion on this point see Chapter 10 here, and also Penrose (1977), and the book Complex Manifold Techniques in Theoretical Physics (Eds.: D. Lerner and P. Sommers; Pitman, 1979).

We shall return to matters of sheaves and cohomology later. Let us now consider more explicitly the sort of condition that must be imposed (and that can be refined and spelled out more explicitly within a sheaf theoretic framework) on a twistor function in order to ensure that the field it generates has positive frequency. It is necessary first to build up some apparatus useful in evaluating contour integrals. The following result—well known, in a slightly disguised form, from elementary complex analysis—is of fundamental utility:

4.5.2 Lemma. Let $\alpha_{A'}$ and $\beta_{A'}$ be a pair of fixed spinors: then the contour integral formula

$$2\pi i(\alpha_{A'}\beta^{A'})^{-1} = \oint(\alpha^{A'}\pi_{A'}\beta^{B'}\pi_{B'})^{-1}\Delta\pi \tag{4.5.3}$$

is valid, where the contour surrounds the pole $\beta^{B'}\pi_{B'} = 0$ once in the positive sense (or, equivalently, surrounds the pole $\alpha^{A'}\pi_{A'} = 0$ once in the negative sense).

Proof. Since the differential form to be integrated in homogeneous of degree zero, $\pi^{A'}$ can be scaled such that in a suitable basis its components are given by

$$\pi^{A'} = (\lambda\ ,\ 1)\quad , \tag{4.5.4}$$

and the associated differential form $\Delta\pi$ (defined in equation 4.3.6) is given by

$$\Delta\pi = -d\lambda\ . \tag{4.5.5}$$

Writing, in the same basis, $\alpha_{A'} = (a,b)$ and $\beta_{A'} = (f,g)$, it is straightforward to see that

$$\oint(\alpha^{A'}\pi_{A'}\beta^{B'}\pi_{B'})^{-1}\Delta\pi = \oint(a\lambda + b)^{-1}(f\lambda + g)^{-1}d\lambda\ . \tag{4.5.6}$$

Thus, taking the contour to surround the pole at $\lambda = -g/f$, elementary calculus of residues shows that the result of the integral is

(4.5.7) $$2\pi i(ag - bf)^{-1} = 2\pi i(\alpha_{A'}\beta^{A'})^{-1} \quad . \quad \square$$

Armed with this lemma, we can examine the twistor function

(4.5.8) $$f(Z^\alpha) = (P_\alpha Z^\alpha Q_\beta Z^\beta)^{-1} \quad ,$$

and see what sort of field it gives rise to. It is worth noticing that if P_α is any fixed dual twistor with spinor parts given by

(4.5.9) $$P_\alpha = (P_A \, , \, P^{A'}) \quad ,$$

then the following formula is valid (cf. equation 3.5.17):

(4.5.10) $$\rho_x P_\alpha Z^\alpha = p^{A'}\pi_{A'} \quad ,$$

where $p^{A'}(x) = ix^{A'A}P_A + P^{A'}$ is the solution of the primed twistor equation (cf. Section 2.5) associated with the dual twistor P_α . Accordingly, we have the identity

(4.5.11) $$\rho_x (P_\alpha Z^\alpha Q_\beta Z^\beta)^{-1} = (p^{A'}\pi_{A'}q^{B'}\pi_{B'})^{-1} \quad ,$$

where $p^{A'}$ and $q^{A'}$ are solutions of the primed twistor equation. Inserting this identity into the contour integral formula

(4.5.12) $$\phi(x) = \oint \rho_x f(Z^\alpha)\Delta\pi \quad ,$$

we can apply Lemma 4.5.2 in order to evaluate the field $\phi(x)$, obtaining:

(4.5.13) $$\phi(x) = 2\pi i(p_{A'}q^{A'})^{-1} \quad .$$

It is straightforward to verify—using the primed twistor equation—that $\phi(x)$ satisfies the wave equation. But what conditions must be imposed in order to ensure that $\phi(x)$ is of positive frequency? A geometrical argument can be employed to give the correct answer. Note that the twistor function $f(Z^\alpha)$ given in (4.5.8) is singular on the plane P_α and on the plane Q_α (recall that by "the plane P_α" we mean the locus in PT given by $P_\alpha Z^\alpha = 0$; cf. Section 2.6). Thus—providing that we stay away from the intersection of P_α and Q_α—the twistor function (4.5.8), when re-

Figure 4.1: An Elementary State

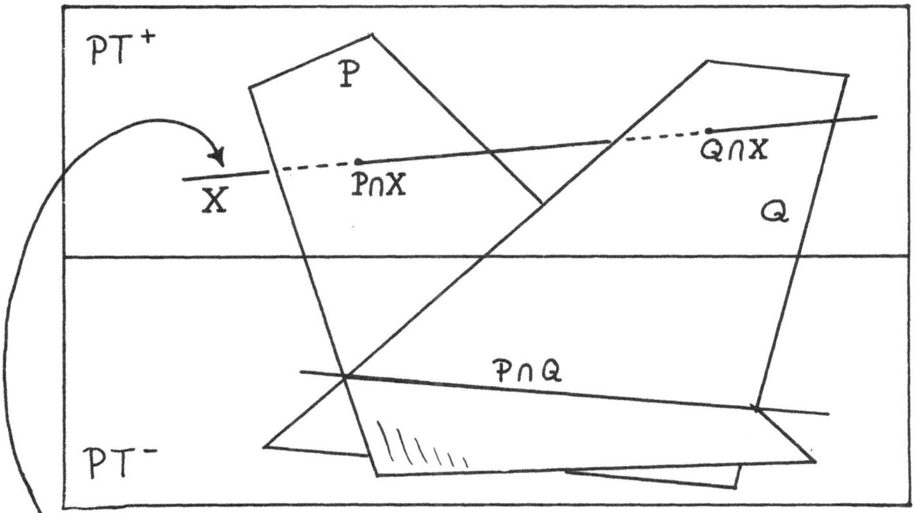

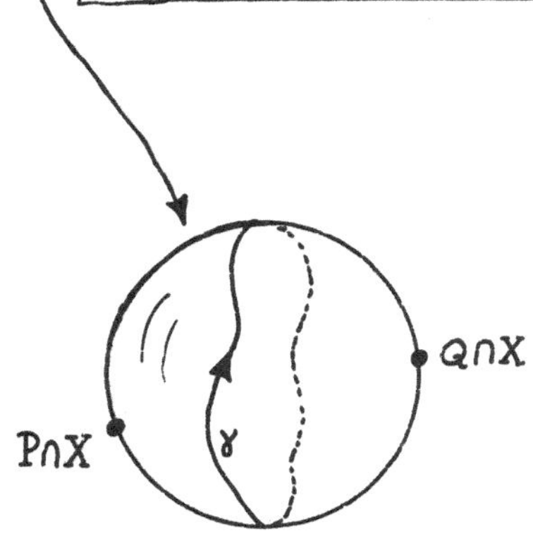

The planes P ($P_\alpha Z^\alpha = 0$) and Q ($Q_\alpha Z^\alpha = 0$) intersect at the line P∩Q ($P_{[\alpha}Q_{\beta]}X^\beta = 0$). Any line X which does not meet the line P∩Q intersects the planes P and Q at a pair of distinct points P∩X and Q∩X. The holomorphic function $f(Z^\alpha) = (P_\alpha Z^\alpha)^{-1}(Q_\alpha Z^\alpha)^{-1}$ when restricted down to the line X is singular only at the points P∩X and Q∩X. Integrating along a contour γ which separates these two points, one obtains a result $\phi(X)$ which depends of course on the choice of line X. Each line X corresponds to some point $x^{AA'}$ in complex Minkowski space, and if X lies in PT^+ then $x^{AA'}$ belongs to the future tube. Since in this case $\phi(X)$ is defined for every line in PT^+, we obtain a field $\phi(x^{AA'})$ which is non-singular throughout the future tube and thus exhibits the positive frequency property.

tricted down to any projective line, will have precisely two singularities: a pole at the intersection of the line with P_α, and another pole at the intersection of the line with Q_α. Accordingly, for every line $X^{\alpha\beta}$ that avoids the line of intersection of the planes P_α and Q_α —i.e. avoids $P_{[\alpha}Q_{\beta]}$ —we have a pair of distinct poles, and a contour exists separating the two poles, and we can evaluate the integral so as to obtain a well-defined value for the field $\phi(x)$. The <u>singularities</u> of the field are at spacetime points corresponding to lines in PT that meet the "bad" line $P_{[\alpha}Q_{\beta]}$. Now we want, for positive frequency, a field that is non-singular throughout $\overline{CM}^+$. Since points in $\overline{CM}^+$ correspond to lines in $\overline{PT}^+$, we accordingly require that every line in $\overline{PT}^+$ should avoid the bad line. The only way to ensure this is to demand that the bad line lie entirely within PT^-. Now if $P_{[\alpha}Q_{\beta]}$ lies in PT^-, then the complex conjugate line $\bar{P}^{[\alpha}\bar{Q}^{\beta]}$ must lie in PT^+. This is ensured if we have

$$P_\alpha \bar{P}^\alpha > 0 \quad , \quad Q_\alpha \bar{Q}^\alpha > 0 \tag{4.5.14}$$

together with

$$P_{[\alpha}Q_{\beta]}\bar{P}^{[\alpha}\bar{Q}^{\beta]} > 0 \quad , \tag{4.5.15}$$

this latter condition being imposed in order that the imaginary part of the complex spacetime point corresponding to $\bar{P}^{[\alpha}\bar{Q}^{\beta]}$ should be <u>timelike</u> (in accordance with equation 2.5.10).

Summing up, we see that the field $\phi(x)$ given by equation (4.5.13), which is generated by the twistor function (4.5.8), will be positive frequency if and only if P_α and Q_α satisfy equations (4.5.14) and (4.5.15), these amounting to the conditions that the planes P_α and Q_α intersect in a line that lies entirely in PT^-. This configuration is illustrated in Figure 4.1.

It may not be evident at a glance that formula (4.5.13) describes a positive frequency field if equations (4.5.14) and (4.5.15) hold, although from the geometric argument above we know that it must. Indeed, in formula (4.5.13) the singularity structure of $\phi(x)$ is not as evident as one would perhaps like. This situation can be remedied with the following observations:

<u>4.5.16 Proposition</u>. The solution $p^{B'}(x)$ of the primed twistor equation, corresponding to the dual twistor P_α, is given by

$$p^{B'}(x) = P_\alpha X^{\alpha B'} \quad , \tag{4.5.17}$$

where $X^{\alpha B'}$ is given as in equation (3.5.16).

The proof is easy, and can be left to the reader. Now, using Proposition 4.5.16, we can rewrite (4.5.13) in the form

$$\phi(x) = 2\pi i(P_\alpha X^\alpha{}_{A'} Q_\beta X^{\beta A'})^{-1} \quad . \tag{4.5.18}$$

Using equation (3.5.20) this becomes

$$\phi(x) = 2\pi i(P_{[\alpha}Q_{\beta]}X^{\alpha\beta})^{-1} \quad , \tag{4.5.19}$$

which, according to equation (2.5.9), gives

$$\phi(x) = \pi i k[(x^a - r^a)(x_a - r_a)]^{-1} \quad , \tag{4.5.20}$$

where r^a is the complex spacetime point corresponding to the bad line $P_{[\alpha}Q_{\beta]}$, and k is the normalization factor $P_\alpha Q_\beta I^{\alpha\beta}$.

Thus we see that $\phi(x)$ is singular whenever x^a is complex null separated from the point r^a. This is consistent with our earlier observation that $\phi(x)$ should be singular precisely on the set of spacetime points corresponding to the set of lines which meet the bad line—since the meeting of a pair of lines correponds to the complex null separation of the associated spacetime points. But so long as $x^a \varepsilon \overline{CM}^+$, the field $\phi(x)$ will not be singular, since a point in $\overline{CM}^+$ can never be null separated from a point in CM^-, as can be verified with a short calculation.

Fields of the form (4.5.20) are called <u>elementary states</u>: they are the prototypes of non-singular, asymptotically well-behaved, normalizable, positive frequency wave functions—and they admit a remarkably simple characterization in twistor terms, by means of twistor functions of the form (4.5.8). The twistor function (4.5.8) illustrates the sort of singularity structure that must be exhibited in order that the associated spacetime field be positive frequency—namely, the singularities of $f(Z^\alpha)$ fall, in $\overline{PT}^+$, into two disjoint open sets: this ensures that the singularities of $f(Z^{\alpha})$ when restricted to any line in $\overline{PT}^{+}$ will fall into two disjoint sets, and thus ensures that on each line a contour integral can be taken, thereby giving a well-defined field throughout $\overline{CM}^{+}$. For further discussion the reader may wish to consult Penrose (1968b), pp. 76-80.

Chapter 4, Notes

1. For a standardized treatment of zero rest mass fields from a spinor point of view the reader may wish to consult Penrose (1965), Pirani (1965), or Penrose (1968a).

2. Strictly speaking, for positive frequency all that is required is that the field be well-defined throughout CM^{+}. In order to be normalizable (in a conventional sense) the wave function must be defined on real Minkowski space as well.

3. This proposition is proved, in essentially the manner described here, in Penrose and MacCallum (1972), section 5.3. The same result, in a somewhat more primitive notation, can be found in Penrose (1968b), section 3, and in Penrose (1969).

CHAPTER 5

TWISTOR QUANTIZATION: MASSIVE FIELDS

5.1 Operators for Momentum and Angular Momentum.

If we are to apply rule (4.2.2) in order to quantize systems composed of one twistor, then for systems composed of several twistors it would seem natural to impose this rule individually for each of the constituent twistors: with this hypothesis the quantization rule for a system Z^α_i of several twistors is

$$\bar{Z}^i_\alpha \longrightarrow \hat{Z}^i_\alpha (= -\partial/\partial Z^\alpha_i) \quad , \tag{5.1.1}$$

and it should be evident that (5.1.1) leads to the following commutation relations:

$$[Z^\alpha_i \, , \, \hat{Z}^j_\beta] = \delta^\alpha_\beta \delta^j_i \quad . \tag{5.1.2}$$

Applying rule (5.1.1) to the classical n-twistor expression for the kinematical twistor, given in equation (3.3.11), one obtains the following expression for the resulting operator:

$$\hat{A}^{\alpha\beta} = 2Z^{(\alpha}_i I^{\beta)\gamma} \hat{Z}^i_\gamma \quad , \tag{5.1.3}$$

and it is straightforward to verify—using the identity $I^{(\alpha\beta)} = 0$ and commutation relations (5.1.2)—that the factor ordering in expression (5.1.3) is irrelevant.

It is tempting to conjecture that the spinor parts of $\hat{A}^{\alpha\beta}$ are the momentum and angular momentum operators for the n-twistor system Z^α_i . And indeed, if we put

$$\hat{A}^{\alpha\beta} = \begin{pmatrix} -2i\hat{\mu}^{AB} & \hat{P}^A{}_{B'} \\ \hat{P}_{A'}{}^B & 0 \end{pmatrix} \qquad \hat{A}_{\alpha\beta} = \begin{pmatrix} 0 & \hat{P}_A{}^{B'} \\ \hat{P}^{A'}{}_B & 2i\hat{\mu}^{A'B'} \end{pmatrix} \tag{5.1.4}$$

where $\hat{A}_{\alpha\beta}$ is the Hermitian conjugate of $\hat{A}^{\alpha\beta}$, defined by

$$\hat{A}_{\alpha\beta} = \hat{Z}^i_{(\alpha} I_{\beta)\gamma} Z^\gamma_i \quad , \tag{5.1.5}$$

then the following result is valid:

<u>5.1.6 Proposition.</u> The momentum operator $\hat{P}^a$ and the angular momentum operator

$\hat{M}^{ab}$ defined by $\hat{M}^{ab} = \hat{\mu}^{AB}\varepsilon^{A'B'} + \hat{\mu}^{A'B'}\varepsilon^{AB}$ satisfy together the correct Poincaré commutation relations (4.3.12).

Proof. Difficult way: using (5.1.2) one can, after some computation, arrive at the following commutation relations:

(5.1.7) $$[\hat{A}^{\alpha\beta}, \hat{A}^{\rho\sigma}] = 2\hat{A}^{\alpha(\rho}{}_{I}{}^{\sigma)\beta} + 2\hat{A}^{\beta(\rho}{}_{I}{}^{\sigma)\alpha}$$

(5.1.8) $$[\hat{A}^{\alpha\beta}, \hat{A}_{\rho\sigma}] = 4\delta^{(\alpha}_{(\rho}{}^{I}{}_{\sigma)\gamma}A^{\beta)\gamma} \quad .$$

Decomposing these relations using (5.1.4), (3.2.1), and (2.3.6), then reassembling them in terms of $\hat{P}^{a}$ and $\hat{M}^{ab}$, the result turns out to be (4.3.12). Easy way: note that $\hat{P}^{a}$ and $\hat{M}^{ab}$ are each linear in the corresponding operators for the constituent ZRM subsystems. On account of (5.1.2), kinematical operators for distinct subsystems always commute: using this fact, one can readily verify that since equations (4.3.12) hold for each of the subsystems individually, they also hold for the total kinematical operators. □

It is worth remarking that if we relax the condition $\hat{Z}^{i}_{\alpha} = -\partial/\partial Z^{\alpha}_{i}$ then equations (5.1.3), (5.1.5), (5.1.7) and (5.1.8) do not necessarily automatically imply (5.1.2)—for example, one could a priori equally well have used anticommutators in (5.1.2) rather than commutators. An interesting problem would be to determine the most general set of commutation relations (or anticommutation relations) imposable on Z^{α}_{i} and $\hat{Z}^{i}_{\alpha}$ such that (5.1.7) and (5.1.8) hold. In what follows we shall continue in adopting the "elementary" commutation relations indicated in equations (5.1.1) and (5.1.2)—with, as it turns out, reasonable justification.

Another way, incidentally, of deriving the correct expressions for the operators of momentum and angular momentum is to examine infinitesimal Poincaré transformations, since the kinematical operators arise as the generators of such transformations. Inasmuch as the Poincaré group is a subgroup of U(2,2), the matter can be approached somewhat more generally by first examining the generators for U(2,2) and then specializing to the Poincaré group. Now an infinitesimal U(2,2) transformation must be of the form $U^{\alpha}_{\beta} = \delta^{\alpha}_{\beta} + i\varepsilon\Lambda^{\alpha}_{\beta}$, where ε is small; the condition of unitarity is:

$$(5.1.9) \qquad U^{\alpha}_{\beta}\bar{U}^{\beta}_{\gamma} = (\delta^{\alpha}_{\beta} + i\varepsilon\Lambda^{\alpha}_{\beta})(\delta^{\beta}_{\gamma} - i\varepsilon\bar{\Lambda}^{\beta}_{\gamma}) = \delta^{\alpha}_{\gamma} \quad ,$$

which, with the neglect of ε^2, asserts that Λ^{α}_{β} is Hermitian. Thus infinitesimal U(2,2) transformations are of the form:

$$(5.1.10) \qquad Z^{\alpha}_{i} \longrightarrow Z^{\alpha}_{i} + i\varepsilon\Lambda^{\alpha}_{\beta}Z^{\beta}_{i} \quad , \quad \Lambda^{\alpha}_{\beta} = \bar{\Lambda}^{\alpha}_{\beta} \quad .$$

Using (5.1.2), this transformation can be rewritten in the following way:

$$(5.1.11) \qquad Z^{\alpha}_{i} \longrightarrow Z^{\alpha}_{i} + i\varepsilon[Z^{\alpha}_{i} , \Lambda^{\sigma}_{\rho}\hat{E}^{\rho}_{\sigma}] \quad ,$$

where $\hat{E}^{\alpha}_{\beta}$ is the operator defined by

$$(5.1.12) \qquad \hat{E}^{\alpha}_{\beta} = Z^{\alpha}_{i}\hat{Z}^{i}_{\beta} \quad .$$

Thus, the components of the operator $\hat{E}^{\alpha}_{\beta}$ are the generators of infinitesimal U(2,2) transformations. Proceeding on to Poincaré transformations, one finds the following lemma of some utility:

<u>5.1.13 Lemma</u>. Infinitesimal Poincaré transformations are of the form

$$(5.1.14) \qquad U^{\alpha}_{\beta} = \delta^{\alpha}_{\beta} + i\varepsilon\Lambda^{\alpha}_{\beta} \quad , \quad \Lambda^{\alpha}_{\beta} = P_{\beta\gamma}I^{\alpha\gamma} + \bar{P}^{\alpha\gamma}I_{\beta\gamma} \quad ,$$

where $P_{\alpha\beta}$ is a symmetric twistor, and $\bar{P}^{\alpha\beta}$ is its complex conjugate.

<u>Proof</u>. As indicated in Section 2.3, Poincaré transformations are SU(2,2) transformations which have the extra property that they preserve the infinity twistors. Thus, Λ^{α}_{β} must be tracefree, and must satisfy:

$$(5.1.15) \qquad (\delta^{\alpha}_{\rho} + i\varepsilon\Lambda^{\alpha}_{\rho})(\delta^{\beta}_{\sigma} + i\varepsilon\Lambda^{\beta}_{\sigma})I^{\rho\sigma} = I^{\alpha\beta} \quad ,$$

which, to first order in ε, is the condition

$$(5.1.16) \qquad \Lambda^{[\alpha}_{\gamma}I^{\beta]\gamma} = 0 \quad .$$

The general solution to (5.1.16), momentarily forgetting that we want Λ^{α}_{β} to be Hermitian, is:

$$(5.1.17) \qquad \Lambda^{\alpha}_{\beta} = P_{\beta\gamma}I^{\alpha\gamma} + Q^{\alpha\gamma}I_{\beta\gamma} \quad ,$$

where $P_{\alpha\beta}$ is a symmetric twistor, and $Q^{\alpha\beta}$ is quite arbitrary. With the imposition of the condition that Λ^{α}_{β} is Hermitian, the desired result follows immediately. □

Using Lemma 5.1.13 it is straightforward to see that infinitesimal Poincaré transformations are of the form (5.1.11), with

(5.1.18) $$\Lambda^{\sigma}_{\rho}\hat{E}^{\rho}_{\sigma} = \frac{1}{2}\,(P_{\alpha\beta}\hat{A}^{\alpha\beta} + \bar{P}^{\alpha\beta}\hat{A}_{\alpha\beta}) \quad ,$$

thus confirming that the kinematical operators arise as the generators of such transformations.

5.2 Contour Integral Formulae for Massive Fields.

In view of the description of ZRM fields in terms of holomorphic functions of a single twistor, it is not unnatural for the case of massive fields to consider holomorphic functions of two or more twistors. In the case of massive fields considerably more scope is offered in the structuring of the associated contour integral formulae, since coefficients of both the $\pi^{A'}_{i}$ type as well as the $\hat{\pi}^{i}_{A}$ type can appear. We shall write

(5.2.1) $$\rho_x f(Z^{\alpha}_{j}) = f(\rho_x Z^{\alpha}_{j}) = f(ix^{AA'}\pi_{A'j}\ ,\ \pi_{A'j})$$

for the restriction of an n-twistor function to the spacetime point $x^{AA'}$. The associated contour integral formula for f(Z) is as follows:

(5.2.2) $$\phi^{A'\ldots i\ldots}_{A\ \ldots j\ldots} = \oint \rho_x \hat{\pi}^{i}_{A}\ \ldots\ \pi^{A'}_{j}\ \ldots\ f(Z)\Delta\pi\ ,$$

where $\Delta\pi$ is the natural projective differential form on the space of the $\pi^{A'}_{i}$ variables, given by:

(5.2.3) $$\Delta\pi = \varepsilon^{ij\ldots k}\varepsilon^{\ell m\ldots n}\pi_{iA'}d\pi^{A'}_{\ell}\,d\pi_{jB'}d\pi^{B'}_{m}\ \ldots\ d\pi_{kC'}d\pi^{C'}_{n} \quad .$$

Note that $\Delta\pi$ is a (2n-1)-form, homogeneous of degree 2n in $\pi^{A'}_{i}$. Also, note that it is invariant under internal SU(n) transformations (i.e. $\pi^{A'}_{i} \longrightarrow U^{j}_{i}\pi^{A'}_{j}$).

The field produced in (5.2.2) has group indices as well as spinor indices. Thus, it corresponds not to a unique particle state, but rather to an entire <u>multiplet</u> of states—moreover, the multiplet can include states of several different spins. The field (5.2.2) corresponds to a definite particle state only if the twistor function $f(Z_i^\alpha)$ is suitably restricted. If $f(Z_i^\alpha)$ is placed in a simultaneous eigenstate of a suitably complete set of commuting operators, then a unique field belonging to the multiplet (5.2.2) will be selected out, with all the others vanishing. Equivalently, if $f(Z_i^\alpha)$ corresponds to a definite particle state, then of the myriad possibilities for spinor coefficient structures appearing in (5.2.2), only one independent combination will lead to an integral which does not vanish. Thus, the twistor particle program—when taken in its most basic form—consists of the analysis of a pair of intertwined problems: first, the construction of suitable sets of holomorphic differential operators, which correspond to quantum mechanical observables, and whose eigenvalues are the quantum numbers of the associated states; and second, the classification of all possible spinor coefficient structures, with the elucidation of the properties of the states which they project out. As stated in Penrose (1977), within the framework of this categorization we shall be able to describe "all the quantum numbers possessed by the elementary particles of Nature"—and moreover, the twistor function "affords a complete description of a particle, both as regards its external and internal parameters".

5.3 The Mass Operator.

The title of this section is something of a misnomer inasmuch as the natural operator which arises in twistor theory is actually the mass-squared operator. In fact, it doesn't seem possible to exhibit a simple twistor operator expression which gives the mass linearly. It is perhaps worth noting that, in view of this fact, the Gell-Mann mass formula for baryons could not be expected to arise in any particularly profound way—and I do not count the "standard" derivation of the Gell-Mann formula, using perturbation theory and group theoretical methods, as being in any sense profound—within the twistor framework. If the Gell-Mann formula were modified so as to apply with squared masses for the baryons, and if higher order correction terms were included so as to account for the discrepancies which would otherwise materialize, then more scope would be offered for the development of hadronic mass formulae using twistor methods. Indeed, such an approach has been advocated by Perjés and Sparling (1976), and shows considerable promise.

Following the pattern used in the proof of Proposition 4.3.7 it is straightforward to see that the momentum operator

$$\hat{P}^{A'A} = \pi^{A'}_{i}\hat{\pi}^{Ai} \tag{5.3.1}$$

satisfies the identity,

$$i\nabla_{AA'}\rho_x = \rho_x\hat{P}_{AA'} \tag{5.3.2}$$

when acting on holomorphic functions of n twistors. And thus for the mass-squared operator

$$\hat{M}^2 := \hat{P}_{AA'}\hat{P}^{AA'} \quad , \tag{5.3.3}$$

we have the identity:

$$\nabla_{AA'}\nabla^{AA'}\rho_x + \rho_x\hat{M}^2 = 0 \quad , \tag{5.3.4}$$

which shows that if $f(Z^{\alpha}_{i})$ is in an eigenstate of the twistor operator $\hat{M}^2$, then the associated spacetime field (henceforth when we say "field" we shall usually mean "spacetime field") will automatically be in the appropriate eigenstate of the conventional spacetime mass-squared operator $-\nabla_{AA'}\nabla^{AA'}$. Observe, incidentally, that using equation (5.3.1) we can rewrite the operator $\hat{M}^2$ as

$$\hat{M}^2 = \hat{M}_{ij}\hat{M}^{ij} \quad , \tag{5.3.5}$$

where $\hat{M}_{ij}$ and $\hat{M}^{ij}$ are <u>partial mass tensor operators</u>, defined by

$$\hat{M}_{ij} = \pi_{A'i}\pi^{A'}_{j} \quad , \quad \hat{M}^{ij} = \hat{\pi}_{Ai}\hat{\pi}^{A}_{j} \quad , \tag{5.3.6}$$

obtained in accordance with the quantization of formulas (3.5.2).

5.4 The Spin Operator.

The momentum operator and the angular momentum operator, along with the mass-squared operator, are all examples of observables which can be expressed directly in twistor terms. It should be noticed that these observables are all differential operators with polynomial coefficients. All of our observables, in fact, belong to the ring of differential operators with polynomial coefficients. Unfortunately, not as much as would perhaps be desired is known about this ring, from a mathematical point of view—indeed, in this connection M.F. Atiyah (1976) remarks, "This is a very interesting non-commutative ring which has been neglected by algebraists: it deserves a lot of study." Within the ring there are certain elements which are Hermitian, and therefore qualified to be regarded as observables. Our problem, then, is to devise a systematic way of physically interpreting these observables, and the states on which they act. Alternatively, given a physical observable taken from the milieu of elementary particle phenomenology, one may seek the appropriate differential operator with polynomial coefficients to which it corresponds. Whether the right operator has been chosen can be determined by examining its eigenvalue spectrum and its commutation relations with other known observables, to see whether these are compatible with the desired results: there is no a priori guarantee, for a given physical observable, that an appropriate twistor operator can be found—but it is implicit in the twistor particle hypothesis that this should, in fact, always be the case.

Let us now, in the light of these remarks, consider the spin operator. A field $\phi_{A\ldots B}$ with mass m is said to be in a definite state of total spin if it is completely symmetric in all its indices; the spin is then one-half the number of indices. If the field has indices of both types, then, in addition to being symmetric with respect to both sets of indices, it must be divergence-free: this ensures that by taking appropriate derivatives the field can be converted—without loss of information—into a field with indices of only one type, completely symmetric. According to relativistic quantum mechanics, the spin-vector operator is defined by

$$\hat{S}_a = \frac{1}{2}\,\varepsilon_{abcd}\hat{M}^{bc}\hat{P}^d \quad , \tag{5.4.1}$$

and the total spin operator is

$$\hat{S}^2 = -\hat{S}_a\hat{S}^a/m^2 \quad , \tag{5.4.2}$$

which has eigenvalues of the form s(s+1), where s is the spin. Substituting the formula $\hat{M}^{ab} = \hat{\mu}^{AB}\varepsilon^{A'B'} + \hat{\mu}^{A'B'}\varepsilon^{AB}$ into equation (5.4.1) we obtain:

(5.4.3) $$\hat{S}^{AA'} = -i\hat{\mu}^{AB}\hat{P}_B^{A'} + i\hat{\mu}^{A'B'}\hat{P}_{B'}^{A} \quad ,$$

and the appropriate twistor operator is generated by putting

(5.4.4) $$\hat{\mu}^{AB} = i\omega_j^{(A}\hat{\pi}^{B)j} \quad , \quad \hat{\mu}^{A'B'} = -i\hat{\omega}^{j(A'}\pi_j^{B')} \quad ,$$

along with (5.3.1). The following result was established by G.A.J. Sparling early in the year 1975:

5.4.5 Proposition. If a twistor function $f(Z_i^\alpha)$ is in an eigenstate of the twistor spin operator $\hat{S}^2$ defined above, with eigenvalue $s(s+1)$, then the field multiplets associated to $f(Z_i^\alpha)$ by means of contour integral formulae of the form (5.2.2) are necessarily of spin s.

The only proof that I am aware of for this result is rather complicated and not particularly illuminating, so it will be omitted. It would be very interesting if someone could devise a short and concise proof, perhaps using cohomological methods—this would, of course, entail understanding the twistor functions $f(Z_i^\alpha)$ much better, perhaps using some of the ideas suggested in Chapter 10.

In addition to beina in an eigenstate of the total spin operator, we may require that the twistor function is in an eigenstate of some component of the spin-vector operator in a particular direction. Thus we define the operator

(5.4.6) $$\hat{S}_z = -z_a\hat{S}^a/m \quad ,$$

where z_a is a unit spacelike vector, orthogonal to the expectation value of the momentum of the state. If the total spin is s , then the eigenvalue s_z of the operator $\hat{S}_z$ lies in the range

(5.4.7) $$-s \le s_z \le s \quad . \qquad [s_z = s (\mathrm{mod}\ 1)]$$

If, for example, $s = 1/2$, then s_z can assume the two values $s_z = \pm 1/2$, corresponding to "spin up" and "spin down". Now if z_a is a unit spacelike vector orthogonal to the momentum P_a of a plane wave solution then there exists a pair of spinors o^A and ι^A such that

(5.4.8) $$z^{AA'} = (o^A \bar{o}^{A'} - \iota^A \bar{\iota}^{A'})/\sqrt{2} \quad ,$$

(5.4.9) $$p^{AA'} = m(o^A \bar{o}^{A'} + \iota^A \bar{\iota}^{A'})/\sqrt{2} \quad ,$$

with $o_A \iota^A = 1$. Moreover, o^A and ι^A are determined completely, up to phase. For a symmetric field $\phi\ldots$ let us define the quantity ϕ_r by

(5.4.10) $$\phi_r = \phi_{AB\ldots CD\ldots} \overbrace{o^A o^B \ldots}^{s-r} \overbrace{\iota^C \iota^D \ldots}^{r} \quad ;$$

that is to say, we contract $\phi\ldots$ with o^A a total of s-r times, and with ι^A a total of r times. Then the field $\phi\ldots$ is in a state of $s_z = s-r_o$ if r_o is the only value of r for which ϕ_r is non-vanishing. Thus, for example, a spinor field ϕ_A is in a state of $s_z = 1/2$ if $\phi_A \iota^A = 0$, and is in a state of $s_z = -1/2$ if $\phi_A o^A = 0$.

5.4.11 Proposition. If a twistor function is in an eigenstate of the operator $\hat{S}_z$, then the associated field multiplets are automatically in a definite state of s_z , in the sense described above.

Again, the proof is a bit tedious and will be omitted. It is perhaps worth mentioning, however—for the sake of the reader who may himself wish to verify Propositions 5.4.5 and 5.4.11—that the attendant calculations simplify considerably if one works with momentum eigenstates.

Summarizing, we see that there exist appropriate twistor operators $\hat{S}^2$ and $\hat{S}_z$ such that if a twistor function is in an eigenstate of these operators, then the associated field—or fields, in the event an entire multiplet is being characterized—will automatically emerge in the correct eigenstate.

5.5 Internal U(n) Casimir Operators.

We now know how to characterize in explicit twistor terms the operators corresponding to momentum, angular momentum, mass, and spin. In addition, we require our twistor functions to be in eigenstates of certain internal obserbables, i.e. observables whose eigenvalues remain unchanged when the states upon which they act are subjected to Poincaré transformations. These observables determine the behavior of twistor functions under internal unitary transformations, i.e. transformations of the form

$$f(Z_i^{\alpha}) \longrightarrow f(U_i^j Z_j^{\alpha}) \quad , \tag{5.5.1}$$

with U_i^j unitary. By requiring $f(Z_i^{\alpha})$ to be in a spin eigenstate, and simultaneously in an eigenstate of the Casimir operators of the internal unitary group, as well as in an eigenstate of the Casimir operators of certain subgroups of the internal unitary group, one can arrange for a unique state to be selected out from the field multiplet (5.2.2) as non-vanishing. The set-up varies according to how many twistors are being considered—we shall consider here the first three cases:

One Twistor. In this case there is only one Casimir operator—namely, the "homogeneity" operator:

$$Z^{\alpha}\partial/\partial Z^{\alpha} = -Z^{\alpha}\hat{Z}_{\alpha} \quad , \tag{5.5.2}$$

which arises as the generator of infinitesimal unitary transformations—i.e., U(1) transformations—acting on Z^{α}. The corresponding observable is the helicity operator:

$$\hat{s} = \frac{1}{2} Z^{\alpha}\hat{Z}_{\alpha} - 1 \quad , \tag{5.5.3}$$

as described in Section 4.2.

Two Twistors. Infinitesimal internal U(2) transformations are generated by the operator

$$E_i^j = Z_i^{\alpha}\hat{Z}_{\alpha}^j \quad , \quad (i = 1, 2) \quad . \tag{5.5.4}$$

There are two Casimir operators for U(2), namely:

$$E_i^i \quad , \quad E^{[i}{}_i E^{j]}{}_j \qquad (i = 1, 2) \quad . \tag{5.5.5}$$

The state can be further characterized by breaking the U(2) symmetry, and looking at the U(1) subgroup acting on the first twistor alone—and the Casimir operator associated with that group is the homogeneity operator,

(5.5.6) $$-Z_1^{\alpha}\hat{Z}^1_{\alpha} \quad .$$

Nothing new is obtained by looking at the U(1) subgroup acting on the second twistor, since if the twistor function is in an eigenstate of $Z_1^{\alpha}\hat{Z}^1_{\alpha}$ and $E^j_j = Z_1^{\alpha}\hat{Z}^1_{\alpha} + Z_2^{\alpha}\hat{Z}^2_{\alpha}$, then it will be in an eigenstate of $Z_2^{\alpha}\hat{Z}^2_{\alpha}$ automatically.

Thus, there are three "new" internal observables for two-twistor systems—the two "homogeneities", and the quadratic operator $E^{[i}_{i}E^{j]}_{j}$. Let us denote by $\tilde{E}^j_i$ the tracefree part of E^j_i , i.e.:

(5.5.7) $$\tilde{E}^j_i = E^j_i - \frac{1}{2} E^k_k \delta^j_i \quad .$$

For some purposes it is useful to consider the quadratic operator

(5.5.8) $$I^2 = \tilde{E}^{[i}_{i}\tilde{E}^{j]}_{j} \quad , \qquad (i = 1,\ 2)$$

which, of course, can if desired be expressed in terms of E^i_i and $E^{[i}_{i}E^{j]}_{j}$. A remarkable degeneracy arises in the case of two twistors:

<u>5.5.9 Proposition</u>. When acting on two-twistor functions that are in mass eigenstates, the following operator identity holds:

(5.5.10) $$\hat{S}^2 = \hat{I}^2 \quad ,$$

where $\hat{S}^2$ is the total spin operator (cf. previous section).

This result was first established early in 1974 by K.P. Tod, the corresponding "classical" result having been obtained earlier by Z. Perjés [cf. his 1975 paper, section IV]. The proof is simply by routine algebra, and can be left to the reader. What is remarkable is that one of our new internal observables turns out to be degenerate with an observable we already had—namely, the spin. Thus, there are really only <u>two</u> "new" observables for two-twistor systems—the two homogeneities.

<u>Three Twistors</u>. In this case, life is more complicated. Infinitesimal U(3) transformations are generated the operator

(5.5.11) $$E^j_i = Z_i^{\alpha}\hat{Z}^j_{\alpha} \quad , \qquad (i = 1,\ 2,\ 3)$$

and there are three Casimir operators, namely:

(5.5.12) $$E^i_i \quad , \quad E^{[i}_i E^{j]}_j \quad , \quad E^{[i}_i E^j_j E^{k]}_k \quad . \quad (i = 1, 2, 3)$$

For additional observables, we can select out the U(2) subgroup acting on the first two twistors. Then, following the discussion of two twistors above, we obtain three more observables:

(5.5.13) $$E^i_i \quad , \quad E^{[i}_i E^{j]}_j \qquad (i = 1, 2)$$

and E^1_1 , making six internal observables altogether. Proposition 5.5.9 is <u>not</u> valid for three-twistor systems, since the spin necessarily involves all three twistors. This is not to say that $\hat{I}^2$ has no meaning for three-twistor systems—indeed, it does: it turns out to be the <u>hadronic</u> <u>isospin</u> operator, as will be discussed in Chapter 6.

Thus for three-twistor systems we have as new internal observables the three homogeneities, the two-twistor "isospin" operator $\hat{I}^2$, and the two SU(3) Casimir operators

(5.5.14) $$\hat{C}_2 = \tilde{E}^{[i}_i \tilde{E}^{j]}_j \quad , \quad \hat{C}_3 = \tilde{E}^{[i}_i \tilde{E}^j_j \tilde{E}^{k]}_k \qquad (i = 1, 2, 3) \ ,$$

where $\tilde{E}^i_j$ is defined by

(5.5.15) $$\tilde{E}^i_j = E^i_j - \frac{1}{3} E^k_k \delta^i_j \qquad (i = 1, 2, 3) \quad .$$

Again, using (5.5.15) it is straightforward to verify that $\hat{C}_2$ and $\hat{C}_3$ can be expressed in terms of the three U(3) Casimir operators given in (5.5.12).

For systems composed of larger numbers of twistors, the analysis proceeds along similar lines—although it must be said that things become quite complicated for systems composed of four or more twistors, and work has only really just begun for such cases. As a "standard" arrangement for n twistors, one may envisage a nested set of unitary groups

(5.5.16) $$U(1) \subset U(2) \subset U(3) \subset \cdots \subset U(n-1) \subset U(n) \quad ,$$

where U(n) acts on all n of the twistors, U(n-1) acts on the first n-1 of these, **and** so on and so forth, down to the group U(1), which only acts on the very first of the n twistors. Now let us consider the set of all the Casimir operators of all these unitary groups. It is a straightforward task to verify that these operators all commute amongst each other. Thus we can take the entire collection of these Casimir operators to be a set of commuting observables for our n-twistor particle state. To this set we may wish to adjoin certain other operators, corresponding, for example, to mass and spin, and in this way proceed to establish a set of observables sufficiently maximal for the purpose of unambiguously identifying the particle state. It should be noted that on account of the fact that the dimensionality of twistor space is four, Casimir operators of order greater than four vanish identically, for any value of n. For $n \geq 4$ the total number of internal unitary group Casimir operators built in accordance with the nested sequence 5.5.16 is 4n-6. For certain purposes it is useful to consider schemes alternative to the "standard" arrangement just described. Examples of various alternative schemes arise, as we shall see, in Chapters 7 and 8, in the context of studying both hadrons and leptons. Alternative schemes must be employed also in the analysis of many-particle states.

CHAPTER 6

THE LOW-LYING BARYONS

6.1 The Quark Model.

Most of the known elementary particles fall into two broad classes: leptons and hadrons. The only particles which fall outside of these classes are the photon, the hypothetical graviton, and the hypothetical "weak" bosons (the W-particles, and the Z-particle). In view of increasing evidence favoring a unification of the weak and electromagnetic interactions, it is not unfair to say that the weak bosons and the photon should "morally" be thrown in with the leptons. Then—ignoring the graviton (whose presence in the world is a bit of a nuisance for particle physicists)—there are two broad classes of particles, i.e. leptons and hadrons, and each class can be subdivided into two subclasses: bosons and fermions. Bosons have integral spin and fermions have half-integral spin. Bosonic hadrons are called "mesons", and fermionic hadrons are called "baryons". In this chapter we shall be concerned with the twistor representation of those baryons that lie on the lower end of the mass spectrum.

Hadrons can be regarded as—at least in some sense—being built up out of certain fundamental units called "quarks". Each baryon, for example, is composed of three quarks. There are several different kinds of quarks—at least three, probably more—and so by choosing various combinations of three quarks, various kinds of baryons are manufactured. Many people like to think of quarks very literally, and regard baryons as being composite particles. In what follows we shall adopt a more conservative (and more reasonable) approach, and regard quarks purely from an abstract viewpoint—according to our view, the quark model merely provides a convenient descriptive language for many of the observed group-theoretical aspects of hadron phenomenology, and the remarks which follow should be interpreted in just that sense[(1)].

The "old" hadrons—in particular, those hadrons known before 1974—can be described in terms of three distinct types of quarks; for many of the post-1974 hadrons, it appears that at least four types of quarks are required. For the moment, let us consider just the first three types of quarks. These quarks are distinguished by varying assignments of quantum numbers. In addition to these three "light" quarks, we also have the three corresponding antiquarks. Each quark state has spin 1/2: the quarks have unprimed spinor indices, and the antiquarks have primed spinor indices. The three quark states will be denoted

(6.1.1) $$\pi^{iA} = (u^A , d^A , s^A) \quad ,$$

and the three antiquarks will be denoted

(6.1.2) $$\pi_i^{A'} = (u^{A'} , d^{A'} , s^{A'}) \quad ,$$

where the symbols u, d, and s stand for "up", "down", and "strange" (or, according to some accounts, "sideways"). The use of the symbol π^{iA} for the quark triplet, and the symbol $\pi_i^{A'}$ for the antiquark triplet, is for the sake of a bit of stylistic augury.

In hadron dynamics there are two quantum numbers which are always strictly conserved—the electric charge and the baryon number; and if a state composed of several quarks is represented by a product of quark states, then the total electric charge and baryon number are obtained by summing the values for the various constituent states. In strong interactions and electromagnetic interactions (but <u>not</u> weak interactions) another quantum number is conserved, which is analogous to electric charge in certain ways, called the <u>hypercharge</u>. Each quark state is assigned a particular electric charge, hypercharge, and baryon number—and, for various reasons which can be justified in many ways, these quantum numbers take on peculiar fractional values:

quark state:	u^A	d^A	s^A	$u^{A'}$	$d^{A'}$	$s^{A'}$
charge	2/3	-1/3	-1/3	-2/3	1/3	1/3
hypercharge	1/3	1/3	-2/3	-1/3	-1/3	2/3
baryon number:	1/3	1/3	1/3	-1/3	-1/3	-1/3

<u>6.1.3 Quark quantum number assignments.</u>

It is convenient to regard the u^A quark and the d^A quark as being two distinct states of an "isospinor" quark state n^{aA} according to the scheme

$$\begin{cases} n^{aA} = (u^A\ ,\ d^A) \\ n_a^{A'} = (u^{A'}\ ,\ d^{A'}) \end{cases} \qquad (a = 1,\ 2) \tag{6.1.4}$$

The indices a,b,... are called isospin indices (not to be confused with Minkowski space indices). A product of various quark isospinors and antiquark isospinors is said to be in a definite state of <u>total isospin</u> if it is tracefree and symmetric with respect to its isospin indices; the total isospin is then one-half the number of free isospin indices present. The strange quark s^A counts as an isoscalar, and does not contribute to the isospin of a state. The isospin eigenvalue is denoted I.

There are observed in nature six distinct types of well-established baryon (B=1) states, distinguished by their hypercharge (Y) and total isospin; these types are labelled as follows:

(6.1.5)

Δ .... Y = 1 , I = 3/2

N .... Y = 1 , I = 1/2

Σ .... Y = 0 , I = 1

Λ .... Y = 0 , I = 0

Ξ .... Y = -1 , I = 1/2

Ω .... Y = -2 , I = 0

For most of these types there are many examples known: type N ("nucleon"), for example, includes the proton and the neutron, as well as many excited states, or "resonances", similar to the proton and neutron inasmuch as they have the same values of the basic quantum numbers B, Y, and I.

Since baryons have B = 1, and quarks have B = 1/3, the simplest way of constructing baryons is out of three quarks. Here we list all possible combinations of three quarks:

$$d^A\, d^B\, d^C \qquad u^A\, d^B\, d^C \qquad u^A\, u^B\, d^C \qquad u^A\, u^B\, u^C$$

$$d^A\, d^B\, s^C \qquad u^A\, d^B\, s^C \qquad u^A\, u^B\, s^C$$

$$d^A\, s^B\, s^C \qquad u^A\, s^B\, s^C$$

$$s^A\, s^B\, s^C$$

6.1.6 Combinations of Three Quarks

These combinations are evidently obtained by considering the "supermultiplet" configuration

(6.1.7) $$\pi^{Ai}\pi^{Bj}\pi^{Ck} \quad ,$$

and allowing the indices i, j, and k to range over the values 1, 2, and 3. By symmetrizing over the spinor indices in each of the combinations in (6.1.6), we obtain a set of ten spin 3/2 states. The quantum numbers of these states are given according to the following scheme:

(6.1.8)
$$\begin{array}{ll} \Delta^- \quad \Delta^o \quad \Delta^+ \quad \Delta^{++} \ \ldots\ldots\ldots\ldots & Y = 1,\ I = 3/2 \\ \Sigma^- \quad \Sigma^o \quad \Sigma^+ \ \ldots\ldots\ldots\ldots & Y = 0,\ I = 1 \\ \Xi^- \quad \Xi^o \ \ldots\ldots\ldots\ldots & Y = -1,\ I = 1/2 \\ \Omega^- \ \ldots\ldots\ldots\ldots & Y = -2,\ I = 0 \quad . \end{array}$$

The superscript on each state denotes the electric charge value. In addition to these ten spin 3/2 states, it is also possible to form eight independent spin 1/2 states from the combinations listed in (6.1.6), the quantum numbers of which are as follows:

(6.1.9)
$$\begin{array}{ll} N^o \quad N^+ \ \ldots\ldots\ldots\ldots & Y = 1,\ I = 1/2 \\ \Sigma^- \quad \Sigma^o\ \Lambda^o \quad \Sigma^+ \ \ldots\ldots\ldots\ldots & Y = 0,\ I_\Sigma = 1,\ I_\Lambda = 0 \\ \Xi^- \quad \Xi^o \ \ldots\ldots\ldots\ldots & Y = 1,\ I = 1/2 \ . \end{array}$$

These eight states comprise the basic "baryon octet", and the various spin 1/2 combinations corresponding to them are given explicitly below:

$$
\begin{array}{lll}
 & d_A d_B u^B \ldots\ldots\ldots\ldots\ldots\ldots\ldots & N^o \text{ (neutron)} \\
 & u_A d_B u^B \ldots\ldots\ldots\ldots\ldots\ldots\ldots & N^+ \text{ (proton)} \\
 & d_A d_B s^B \ldots\ldots\ldots\ldots\ldots\ldots\ldots & \Sigma^- \\
 & u_{(A} d_{B)} s^B \ldots\ldots\ldots\ldots\ldots\ldots\ldots & \Sigma^o \\
(6.1.10) & u_A u_B s^B \ldots\ldots\ldots\ldots\ldots\ldots\ldots & \Sigma^+ \\
 & s_A u_B d^B \ldots\ldots\ldots\ldots\ldots\ldots\ldots & \Lambda^o \\
 & s_A s_B d^B \ldots\ldots\ldots\ldots\ldots\ldots\ldots & \Xi^- \\
 & s_A s_B u^B \ldots\ldots\ldots\ldots\ldots\ldots\ldots & \Xi^o
\end{array}
$$

Note that there are two linearly independent ways of reducing the spinor combination $u^A d^B s^C$ down to a spin 1/2 state: one of these gives the I = 1 state Σ^o , and the other gives the I = 0 state Λ^o . The isospin multiplet content of (6.1.9) and (6.1.10) can be recorded more explicitly as follows:

$$
\begin{array}{lll}
 & n^a_A \, n^b_B \, n^{cB} \, \varepsilon_{bc} \ldots\ldots\ldots\ldots\ldots\ldots\ldots & N \text{ (isodoublet)} \\
 & n^{(a}_{(A} n^{b)}_{B)} \, s^B \ldots\ldots\ldots\ldots\ldots\ldots\ldots & \Sigma \text{ (isotriplet)} \\
(6.1.11) & n^a_B \, s^B \, s_A \ldots\ldots\ldots\ldots\ldots\ldots\ldots & \Xi \text{ (isodoublet)} \\
 & n^a_B \, n^{bB} \, s_A \, \varepsilon_{ab} \ldots\ldots\ldots\ldots\ldots\ldots\ldots & \Lambda \text{ (isosinglet)}
\end{array}
$$

where ε_{ab} is the antisymmetric isospin "epsilon" tensor.

Summing up, we see that from the supermultiplet configuration (6.1.7) we obtain a spin 1/2 octet, and a spin 3/2 decimet—and it is indeed truly remarkable (this fact is perhaps the true essence of the role of SU(3) in physics) that the lowest lying baryon states group themselves naturally into a spin 1/2 octet and a spin 3/2 decimet. Some of the basic physical properties of these states are summarized in Table 6.I. It is important to notice that all of the octet members are stable to strong decays, and are unstable (with the exception of the proton, which is completely stable) only to weak/electromagnetic decay. Amongst the decimet mem-

Table 6.I

The Low-Lying Baryons

Particle	Mass (MeV)	Mean Life or Full Width	Decay Modes	Fraction
p	938.2796(27)	stable	-	-
n	939.5731(27)	918(14) sec	$pe^-\bar{\nu}$	100%
Λ	115.60(05)	$2.632(20)\times10^{-10}$ sec	$p\pi^-$	64.2
			$n\pi^o$	35.8
			$pe^-\nu$	$8.07(28)\times10^{-4}$
			$p\mu^-\nu$	$1.57(35)\times10^{-4}$
			$p\pi^-\gamma$	$.85(14)\times10^{-3}$
Σ^+	1189.37(06)	$.802(5)\times10^{-10}$ sec	$p\pi^o$	51.6
			$n\pi^+$	48.4
			$p\gamma$	$1.24(18)\times10^{-3}$
			$n\pi^+\gamma$	$.93(10)\times10^{-3}$
			$\Lambda e^+\gamma$	$2.02(47)\times10^{-5}$
Σ^o	1192.47(08)	$5.8(1.3)\times10^{-20}$ sec	$\Lambda\gamma$	~100%
			Λe^+e^-	5.45×10^{-3}
Σ^-	1197.35(06)	$1.483(15)\times10^{-10}$ sec	$n\pi^-$	~100%
			$ne^-\nu$	$1.08(04)\times10^{-3}$
			$n\mu^-\nu$	$.45(04)\times10^{-3}$
			$\Lambda e^-\nu$	$.60(06)\times10^{-4}$
			$n\pi^-\gamma$	$4.6(6)\times10^{-4}$
Ξ^o	1314.9(6)	$2.90(10)\times10^{-10}$ sec	$\Lambda\pi^o$	~100%
			$\Lambda\gamma$	0.5±0.5%
Ξ^-	1321.32(13)	$1.654(21)\times10^{-10}$ sec	$\Lambda\pi^-$	~100%
			$\Lambda e^-\nu$	$0.69(18)\times10^{-3}$
			$\Lambda\mu^-\nu$	$(3.5\pm3.5)\times10^{-4}$
Δ(1232)	1230-1234	110-120 MeV	$N\pi$	~99.4
Σ(1385)	[+]1382.3(4)	35(2) MeV	$\Lambda\pi$	88±2
	[0]1382.0±2.5	~35 MeV	$\Sigma\pi$	12±2
	[-]1387.5(6)	40(2) MeV		
Ξ(1530)	[0]1531.8(3)	9.1(5) MeV	$\Xi\pi$	100%
	[-]1535.0(6)	10.1±1.9 MeV		
Ω(1672)	1672.2(4)	$1.1^{+0.4}_{-0.3}\times10^{-10}$ sec	$\Xi^o\pi^-$	100%
			$\Xi^-\pi^o$	
			ΛK^-	

bers, only the Ω^- particle (whose existence was predicted by Gell-Mann on the basis of SU(3) theory, and subsequently confirmed with great drama) is stable against strong decay, and the remaining states are observed as resonances in various strong interactions. Aside from the basic octet and decimet members listed in Table 6.I many additional baryon states have been observed. These also seem to clump together naturally into octet and decimet configurations (as well as some possible Λ singlet states), as will be discussed in Chapter 7.

6.2 The Three-Twistor Model for Low-Lying Baryons.

The transition from the quark model to twistor theory is achieved by interpreting the "quark configuration structure" of a hadron state as the "spinor coefficient structure" which appears in the contour integral formula (5.2.2) relating twistor functions to field multiplets. According to this view the operator $\hat{\pi}^{iA}$ is interpreted as a quark triplet, and we interpret $\pi_i^{A'}$ as an antiquark triplet (although this latter interpretation must—as we shall see—be suitably qualified). Thus the fields constituting the low-lying baryon supermultiplet are given by the following contour integral formula:

$$\phi^{ijk}_{ABC} = \oint \rho_x \hat{\pi}^i_A \hat{\pi}^j_B \hat{\pi}^k_C \, f(Z)\Delta\pi \qquad (i = 1,2,3) \quad . \tag{6.2.1}$$

If f(Z) is in a definite quantum state, then a particular member of the supermultiplet ϕ^{ijk}_{ABC} will be picked out uniquely as being non-vanishing. For example, if f(Z) should be a proton state, then—selecting the correct spinor coefficient structure using (6.1.10)—the only non-vanishing field will be given as follows:

$$\phi_A = \oint \rho_x \hat{u}_A \hat{d}_B \hat{u}^B \, f(Z)\Delta\pi \quad , \tag{6.2.2}$$

where $\hat{u}_A$, $\hat{d}_A$, and $\hat{s}_A$ are the three components of the operator $\hat{\pi}^i_A$.

6.3 Electric Charge, Hypercharge, Baryon Number, and Isospin.

What remains to be shown is how to construct in explicit twistor terms the various hadronic observables which we require f(Z) to be put into an eigenstate of. For our twistor triplet Z^α_i (i = 1,2,3) we shall write

(6.3.1) $$Z_i^\alpha = (U^\alpha , D^\alpha , S^\alpha) \quad ,$$

and analogously the three operators $\hat{Z}_\alpha^i$ (= $-\partial/\partial Z_i^\alpha$) will be labelled according to the scheme

(6.3.2) $$\hat{Z}_\alpha^i = (\hat{U}_\alpha , \hat{D}_\alpha , \hat{S}_\alpha) \quad ,$$

with:

(6.3.3) $$\hat{U}_\alpha = -\partial/\partial U^\alpha \qquad \hat{D}_\alpha = -\partial/\partial D^\alpha \qquad \hat{S}_\alpha = -\partial/\partial S^\alpha \quad .$$

The three twistor operators

(6.3.4) $$\hat{u} = -U^\alpha \hat{U}_\alpha + 2 \qquad \hat{d} = -D^\alpha \hat{D}_\alpha + 2 \qquad \hat{s} = -S^\alpha \hat{S}_\alpha + 2$$

are the "total occupation numbers" for the three types of quarks— e.g., $\hat{u}$ measures the number of u-quarks minus the number of u-antiquarks. The baryon number, hypercharge, and electric charge are then given by :

(6.3.5) $$\hat{B} = \frac{1}{3} (\hat{u} + \hat{d} + \hat{s})$$

(6.3.6) $$\hat{Y} = \frac{1}{3} (\hat{u} + \hat{d} - 2\hat{s}) = -\hat{s} + \hat{B}$$

(6.3.7) $$\hat{Q} = \frac{1}{3} (2\hat{u} - \hat{d} - \hat{s}) = \hat{u} - \hat{B} \ .$$

The three generators for infinitesimal isospin transformations—i.e., SU(2) transformations applied to U^α and D^α —are as follows:

(6.3.8) $$\begin{cases} \hat{I}_1 = -\frac{1}{2} (U^\alpha \hat{D}_\alpha + D^\alpha \hat{U}_\alpha) \\ \hat{I}_2 = -\frac{i}{2} (U^\alpha \hat{D}_\alpha - D^\alpha \hat{U}_\alpha) \\ \hat{I}_3 = -\frac{1}{2} (U^\alpha \hat{U}_\alpha - D^\alpha \hat{D}_\alpha) \quad , \end{cases}$$

and the total isospin operator $\hat{I}^2$ is defined by

(6.3.9) $$\hat{I}^2 = (\hat{I}_1)^2 + (\hat{I}_2)^2 + (\hat{I}_3)^2 \quad .$$

The operators $\hat{I}_1$, $\hat{I}_2$, and $\hat{I}_3$ are the twistorial operator analogues of the three <u>Pauli matrices</u>. It is not difficult, incidentally, to verify that formula (6.3.9) agrees with equation (5.5.8). Note that $\hat{I}_3$ can be written, using (6.2.6), in the form

(6.3.10) $$\hat{I}_3 = \frac{1}{2}(\hat{u} - \hat{d}) \quad ,$$

whence, after some elementary algebra, we derive the famous Gell-Mann Nishijima relation,

(6.3.11) $$\hat{Q} = \hat{I}_3 + \hat{Y}/2 \quad .$$

6.4 Mass and Spin for Three-Twistor Systems.

In the case of three twistors the operator $\hat{M}^{ij}$ only has three independent components, and thus it is convenient to introduce, in this case, an operator $\hat{M}_i$ defined by

(6.4.1) $$\hat{M}_i = \frac{1}{2}\varepsilon_{ijk}\hat{M}^{jk} \quad ;$$

and similarly, we define

(6.4.2) $$M^i = \frac{1}{2}\varepsilon^{ijk}M_{jk} \quad .$$

Using equation (5.3.5) it is then an elementary exercise in SU(3) algebra to verify that the mass-squared operator $\hat{M}^2$ is

(6.4.3) $$\hat{M}^2 = 2\hat{M}_i M^i \quad .$$

Thus the mass-squared operator is given by a sum of three partial mass-squared operators, each of which is the mass-squared operator for one of the three two-twistor subsystems.

The operators $\hat{M}_i$ and M^i also figure into the three-twistor expression for the spin operator. If, as was done in Section 5.5, we denote the tracefree SU(3) generators by $\tilde{E}^j_i$, i.e., we put

(6.4.4) $$\tilde{E}^j_i = Z^\alpha_i \hat{Z}^j_\alpha - \frac{1}{3}\delta^j_i Z^\alpha_k \hat{Z}^k_\alpha \quad ,$$

then the following expression—which was first formulated explicitly by G.A.J. Sparling, in 1975—can be derived for the total spin operator:

(6.4.5) $$\hat{S}^2 = \frac{1}{4}\hat{B}^2 + \frac{3}{2}\hat{B} + \frac{1}{2}\tilde{E}^i_j\tilde{E}^j_i - 2m^{-2}M^i\hat{M}_j\tilde{E}^j_k\tilde{E}^k_i \quad .$$

It is worth noting, incidentally, that the baryon number operator $\hat{B}$, defined by

(6.4.6) $$\hat{B} = -\frac{1}{3}\tilde{E}^i_i + 2 \quad ,$$

with $E^j_i = Z^\alpha_i\hat{Z}^j_\alpha$, can, if one desires, be expressed in terms of the <u>tracefree</u> generators $\tilde{E}^j_i$ with the help of the following rather peculiar result:

<u>6.4.7 Theorem (Perjés)</u>. The baryon number operator $\hat{B}$, when acting on mass-eigenstates, can be expressed in the alternative form,

(6.4.8) $$\hat{B} = 2m^{-2}M^i\hat{M}_j\tilde{E}^j_i + 2 \quad .$$

<u>Proof</u>. Let us first demonstrate that

(6.4.9) $$M^i\hat{M}_jE^j_i = M^i\hat{M}_jZ^\alpha_i\hat{Z}^j_\alpha = 0 \quad .$$

To see this, note that $M^iZ^\alpha_i$ satisfies

(6.4.10) $$M^iZ^\alpha_iI_{\alpha\beta} = 0 \quad ,$$

since three π's are being skewed over in (6.4.10). Similarly, one verifies that

(6.4.11) $$\hat{M}_i\hat{Z}^i_\alpha I^{\alpha\beta} = 0.$$

Equation (6.4.9) follows from equations (6.4.10) and (6.4.11) at once, since if a pair of quantities are annihilated, respectively, by $I_{\alpha\beta}$ and $I^{\alpha\beta}$, then they must necessarily annihilate each other. Substituting $E^j_i = \tilde{E}^j_i + \frac{1}{3}\delta^j_iE^k_k$ into equation (6.4.9), the desired result follows immediately. □

Theorem (6.4.7) is really a bit of an aside, since it has to do more with baryon number than with mass or spin. It does show us, however, that $\hat{S}^2$ can be expressed entirely in terms of the operators $\hat{M}_i$, M^i , and $\tilde{E}^j_i$. Since the baryon number operator is also built up out of these operators, it should not come as a total surprise that a certain connection automatically holds between $\hat{B}$ and $\hat{S}$:

6.4.12 Theorem. If a twistor function $f(Z_i^\alpha)$ is in an eigenstate of $\hat{B}$ and $\hat{S}$, then the eigenvalues B and S must satisfy:

$$3B - 2S = 0 \pmod 2 \quad . \tag{6.4.13}$$

Proof. This result is most easily established if one examines the spinor-coefficient structure associated with $f(Z_i^\alpha)$. The baryon number is $\frac{1}{3}(\hat{\pi}-\pi)$, where $\hat{\pi}$ denotes the number of $\hat{\pi}$-coefficients and π denotes the number of π-coefficients. The spin, on the other hand, must be of the form $\frac{1}{2}(\hat{\pi}+\pi)$ [mod 1], and in fact must lie in the range $0 \leq S \leq \frac{1}{2}(\hat{\pi}+\pi)$. These results immediately imply (6.4.13). □

Equation (6.4.13) says that particles with 3B odd are fermions, and 3B even are bosons. This result is, of course, consistent with the way things work in the real world. It should be stressed that equation (6.4.13) follows quite trivially if one assumes the quark model. Here, however, we have not assumed the quark model (at least in the literal sense) and consequently the relation $3B = 2S \pmod 2$ is a result of a somewhat less trivial character.

6.5 The SU(3) Casimir Operators.

In order to belong to a definite SU(3) multiplet a twistor function $f(Z_i^\alpha)$ must be in an eigenstate of the two Casimir operators $\hat{C}_2$ and $\hat{C}_3$ defined in relations (5.5.14). If $f(Z_i^\alpha)$ is indeed in such an eigenstate, then the associated field multiplet will be an irreducible SU(3) tensor. It should be noted that irreducible SU(3) tensors can—after a judicious application of SU(3) epsilons—always be expressed, in a canonical way, as tracefree tensors which are symmetric both on their upstairs indices as well as their downstairs indices. Irreducible SU(3) multiplets can, accordingly, be labelled by a pair of integers $\{\lambda,\mu\}$ giving the number of indices of each type. Now the Casimir operator $\hat{C}_2$ and $\hat{C}_3$ are themselves rather awkward to work with in practice. Their eigenvalues, however, can be expressed very simply in terms of λ and μ, according to the following formulae:

$$(6.5.1)\qquad \begin{cases} C_2 = \frac{1}{3}(\lambda^2 + \mu^2 + \lambda\mu + 3\lambda + 3\mu) \\ C_3 = \frac{1}{27}(\lambda-\mu)(2\lambda+\mu+3)(\lambda+2\mu+3) \quad . \end{cases}$$

The dimension of the irreducible multiplet $\{\lambda,\mu\}$ is given by the formula

$$(6.5.2)\qquad \dim\{\lambda,\mu\} = \frac{1}{2}(\lambda+1)(\mu+1)(\lambda+\mu+2) \quad ,$$

and one can easily check that $\{1,1\}$ gives an octet, $\{3,0\}$ and $\{0,3\}$ give decimets, $\{2,2\}$ gives a 27-plet, and so on. It would be amusing to find a pair of operators whose eigenvalues gave λ and μ directly.

Since $\hat{C}_2$ and $\hat{C}_3$ are built up out of the tracefree twistor internal SU(3) generators, one might anticipate that certain relationships must necessarily hold between λ, μ, B, and S. These are given as follows:

<u>6.5.3 Theorem (Perjés and Sparling)</u>. Let us define the quantities α, β, and γ by:

$$(6.5.4)\qquad \alpha = \frac{2\lambda+\mu+3}{3} \qquad \beta = \frac{\lambda+2\mu+3}{3} \qquad \gamma = \frac{\lambda-\mu}{3} \quad .$$

Then the following inequalities must hold:

$$(6.5.5)\qquad \begin{cases} (\frac{1}{2}B - \alpha \pm \frac{1}{2})^2 \geq (S + \frac{1}{2})^2 \\ (\frac{1}{2}B + \beta \pm \frac{1}{2})^2 \geq (S + \frac{1}{2})^2 \\ (\frac{1}{2}B + \gamma \pm \frac{1}{2})^2 \leq (S + \frac{1}{2})^2 \quad . \end{cases}$$

Although its proof (which will be omitted) is rather long and involved, this result can be verified quite easily by examining the spinor coefficient structures associated with various multiplets. These inequalities impose severe limitations on the spectrum of allowable multiplets available within a three-twistor scheme. In many cases, if B and $\{\lambda,\mu\}$ are specified, then (6.5.5) will in fact determine S uniquely! For example, for a B = 1 octet (with $\{\lambda,\mu\} = \{1,1\}$) one obtains $\alpha = 2$, $\beta = 2$, $\gamma = 0$; and after a short calculation one deduces from (6.5.5) that S = 1/2.

We list below the results of several such calculations:

(6.5.6)

B	$\{\lambda,\mu\}$	dim$\{\lambda,\mu\}$	S
1	{1,1}	$\underline{8}$	1/2
1	{3,0}	$\underline{10}$	3/2
1	{4,1}	$\underline{35}$	3/2 or 5/2
0	{0,0}	$\underline{1}$	0
0	{1,1}	$\underline{8}$	0 or 1
0	{2,2}	$\underline{27}$	0, 1, or 2
1/3	{1,0}	$\underline{3}$	1/2

These results show that our three-twistor scheme, although adequate for the description of low-lying baryons, will not suffice for a general description of hadrons—this is simply because one does, in fact, observe in nature various hadronic multiplets whose quantum numbers are incompatible with those listed above. One observes baryon octets with $S > 1/2$; one observes baryon decimets with $S > 3/2$; one observes baryon singlet states (which are not allowed in a three-twistor scheme); one observes an octet of $S = 2$ mesons; and one observes (as part of a mixed state) an $S = 1$ meson singlet: none of the states just mentioned are compatible with a description based entirely on three twistors. Accordingly, our framework must be generalized so as to account for these additional states. In the next chapter we shall outline an extended scheme (based of functions of six twistors) which exhibits a flexibility sufficient to enable it to account for a greater variety of hadronic states—in particular, it admits the various states mentioned above which are ruled out when only three twistors are considered.

6.6 The Absence of Color Degrees of Freedom.

One point which needs to be stressed here, which has not been mentioned yet, concerns the role of "color" SU(3) in hadronic structure. In the standard "naive" quark model, one assumes that baryons are effectively bound states of three quarks. Now, regardless of the nature of the forces binding the quarks together, the requirements of Fermi statistics demand that the quarks have in addition to their flavor degrees of freedom (viz.: up, down, and strange ...) also three colors' at their disposal. Thus, according to the color hypothesis there are nine distinct kinds of light quarks—each labelled by a color and a flavor. Within a baryon the quarks are put into a color singlet state, i.e., into a state which is totally antisymmetric with respect to its color SU(3) indices. With this assignment baryons then automatically possess the correct flavor and spin symmetries (i.e., totally symmetric with respect to clumped flavor SU(3) and spin indices). The color hypothesis can then be taken one step further. An octet of color SU(3) bosons (called "colored gluons") is introduced, and it is hypothesized that the subhadronic quark binding forces are due to the exchange of virtual gluons—the resulting theory goes by the name of "quantum chromodynamics". There is a certain amount of evidence in favor of QCD, but this evidence rests on such a plethora of assumptions that—to the critical eye—it is not very convincing. The theory's chief merit is its elegance and its aesthetic simplicity.

Within twistor theory baryons are not assumed to be in any literal sense built up out of bound states of quarks, and consequently the color hypothesis is unnecessary. Of course, if there are no color degrees of freedom, then there are no colored gluons—and thus it is not obvious at all how one might begin to formulate a theory of strong interactions in twistor terms. There are several routes that might be followed towards this end. One approach would be to study twistor diagrams[(2)], or appropriate generalizations thereof, and try to build up reasonable expressions for hadronic scattering amplitudes. In this connection one would inevitably anticipate links with Regge theory. Another route to take, perhaps of a more speculative character, would involve looking at deformations[(3)] of the complex analytic structure of the space of three twistors (or, as it may turn out, suitably related higher dimensional spaces). Although it is not at all evident how one would go about describing strong interaction phenomena in terms of such deformations, the utility of such an approach has been demonstrated admirably in a variety of non-linear problems (Penrose 1976; Ward 1977a and 1977b; Atiyah and Ward 1977; Atiyah, Hitchin, and Singer 1977; Hartshorne 1978; etc.) and it is not unreasonable to propose that hadronic interactions might be amenable to treatment by means of this sort. Finally, any approach to strong interaction physics requires a detailed knowledge of the "internal" geometry of hadrons. In Chapter 10 some of the material necessary towards this end is presented; but clearly, in addition to this, knowledge of a much more specific character is needed. It is worth noting that an operator analogue of the center of mass twistor described in Section 3.5 can be constructed (Hughston and Sheppard, 1979), and in addition to the center of mass operator for the three-twistor system as a whole, in the case of hadrons, we also have three "partial" center of mass operators constructed from the three two-twistor subsystems. It is not unlikely that these operators should play a significant part in understanding various aspects of the structure of hadrons. In particular, the role of the center of mass operator in determining the properties of the magnetic moments of hadrons now seems to be firmly established.

Chapter 6, Notes

1. For standard discussions of the quark model and SU(3) see, for example, Gell-Mann and Ne'eman (1964), Dalitz (1966), Feld (1969), and Feynman (1972).

2. Twistor diagrams were introduced in Penrose and MacCallum (1972), and are discussed at length in Penrose (1975a, pp. 330-369). For additional discussion see, for example, Sparling (1974), Sparling (1975), Hodges (1975), Harris (1975), Ryman (1975), and Huggett (1976). A number of articles on twistor diagrams have been written by A.P. Hodges for Twistor Newsletter, and in the same reference one can find an article by S.A. Huggett and M.L. Ginsberg discussing the cohomological interpretation of certain classes of twistor diagrams. In Popovich (1978) one finds a good summary of many of the heuristic aspects of the analysis of twistor diagrams for hadronic, leptonic, and semileptonic processes. Although we shall not be entering into a discussion of the matter here, it is perhaps worth noting that there exist a number of interesting formal correspondences between twistor diagrams and duality diagrams. A useful reference on dual theory is Jacob (1974). Basic

references to duality diagrams include Harari (1969), Rosner (1969), Neville (1969), and Matsuoka et al (1969). Higher order duality diagrams, which also fit into the twistor framework [where "quark loops" correspond to "helicity flux loops" in twistor diagrams], are discussed in Kikkawa et al (1969). There is something very curious and combinatorial about the theory of duality diagrams, suggestive of some of the principles involved in spin-network theory [Penrose 1971a and 1971b; also see the Twistor Newsletter articles on spin-networks by S.A. Huggett and J.P. Moussouris], and more investigation in this area is certainly called for.

3. Standard references for the theory of deformations of complex analytic structures include Kodaira and Spencer (1958), Kodaira and Spencer (1960), and Morrow and Kodaira (1971). It is first suggested in Penrose (1968b) that gravitation is in some sense due to a shift in the complex analytic structure of twistor space.

CHAPTER 7

MESONS, RESONANCES, AND BOUND STATES

7.1 The Low-Lying Mesons.

Among the observed low-lying meson states two nonets stand out as particularly striking. These include a spin zero nonet of negative intrinsic parity, and a spin one nonet of negative intrinsic parity. At the level of the naive quark model, these nonets can be represented by quark-antiquark pairs. Since quarks and antiquarks are both spin 1/2, pairs of such states can be either spin 0 or spin 1, assuming no orbital angular momentum (cf. Section 7.4) between the quarks. In Table 7.I one finds a list of the relevant states comprising these nonets, together with the hypothesized quark structure for each case:

Table 7.I

The Quark Structure of the Low-Lying Mesons

π^+	$u^A\bar{d}_A$	ρ^+	$u^{(A}\bar{d}^{B)}$
π^o	$u^A\bar{u}_A - d^A\bar{d}_A$	ρ^o	$u^{(A}\bar{u}^{B)} - d^{(A}\bar{d}^{B)}$
π^-	$d^A\bar{u}_A$	ρ^-	$d^{(A}\bar{u}^{B)}$
K^+	$u^A\bar{s}_A$	K^{+*}	$u^{(A}\bar{s}^{B)}$
K^o	$d^A\bar{s}_A$	K^{o*}	$d^{(A}\bar{s}^{B)}$
$\bar{K}^o$	$s^A\bar{d}_A$	$\bar{K}^{o*}$	$s^{(A}\bar{d}^{B)}$
K^-	$s^A\bar{u}_A$	K^{-*}	$s^{(A}\bar{u}^{B)}$
η	$u^A\bar{u}_A + d^A\bar{d}_A - 2s^A\bar{s}_a$	ω	$u^{(A}\bar{u}^{B)} + d^{(A}\bar{d}^{B)}$
η'	$u^A\bar{u}_A + d^A\bar{d}_A + s^A\bar{s}_A$	ϕ	$s^{(A}\bar{s}^{B)}$

Each nonet has a pair of I = 0 members. For the spin 0^- nonet we have η and η', and for the spin 1^- nonet we have ω and ϕ. This is because each nonet is composed of an SU(3) octet and an SU(3) singlet. The octet has an I = 0 state, and the singlet has I = 0; this makes for two I = 0 states altogether. It is possible that the observed I = 0 states are mixtures of pure octet and singlet components. No one knows what the precise principles are which govern the phenomenon of mixing: therefore, the quark structures for these states as listed in Table 7.I are to some extent ad hoc. If we write

$$(7.1.1)\qquad \begin{cases} \eta = \cos\theta\{8\} + \sin\theta\{1\} \\ \eta' = -\sin\theta\{8\} + \cos\theta\{1\} \end{cases}$$

and

$$(7.1.2)\qquad \begin{cases} \phi = \cos\theta\{8\} + \sin\theta\{1\} \\ \omega = -\sin\theta\{8\} + \cos\theta\{1\} \end{cases}$$

and assume the Gell-Mann mass formula (cf., however, Section 5.3) then we obtain the following values for the mixing angles:

(7.1.3)

	θ_{lin}	θ_{quad}
η, η'	-24 ± 1°	-11 ± 1°
ϕ, ω	38 ± 1°	40 ± 1°

where θ_{lin} is the angle obtained if Gell-Mann formula is assumed to be linear in the meson masses, and θ_{quad} is the result obtained in the quadratic case. Nature is being very elusive about the whole matter. In Table 7.II a number of the basic properties of the low-lying mesons are summarized. In most cases only those decay modes are listed for which a definite lower bound on the fraction is known. For further information the reader should consult the most recent tables compiled by the Particle Data Group. The data here comes from the 1978 lists.

Table 7.II: Properties of the Low-Lying Mesons

Particle	Mass	Mean Life or Full Width	Decay Modes	Fraction
$\pi^{\pm}$	139.5669(12)	$2.6030(23)\times10^{-8}$ sec	$\mu^+\nu$	~100%
			$e^+\nu$	$1.267(23)\times10^{-4}$
			$\mu^+\nu\gamma$	$1.24(25)\times10^{-4}$
			$e^+\nu\pi^o$	$1.02(07)\times10^{-8}$
			$e^+\nu\gamma$	$2.15(50)\times10^{-8}$
π^o	134.9626(39)	0.828×10^{-16} sec	$\gamma\gamma$	98.85(05)
			γe^+e^-	1.15(05)
			$e^+e^-e^+e^-$	3.32×10^{-5}
η	548.8(6)	0.85(12)keV	$\gamma\gamma$	38.0
			$\pi^o\gamma\gamma$	3.1
			$3\pi^o$	29.9
			$\pi^+\pi^-\pi^o$	23.6(6)
			$\pi^+\pi^-\gamma$	4.89(13)
			$e^+e^-\gamma$	0.50(12)
			$e^+e^-\pi^+\pi^-$	0.1(1)
			$\mu^+\mu^-$	$2.2(8)\times10^{-5}$
$K^{\pm}$	493.668(18)	$1.2371(26)\times10^{-8}$ sec	$\mu^+\gamma$	63.50(16)
			$\pi^+\pi^o$	21.16(15)
			$\pi^+\pi^+\pi^-$	5.59(03)
			$\pi^+\pi^o\pi^o$	1.73(05)
			$\mu^+\nu\pi^o$	3.20(09)
			$e^+\nu\pi^o$	4.82(05)
			$\mu^+\nu\gamma$	$5.8(3.5)\times10^{-3}$
			$e^+\nu\pi^o\pi^o$	$1.8(+2.4)(-0.6)\times10^{-5}$
			$e^+\nu\pi^+\pi^-$	$3.90(15)\times10^{-5}$
			$\mu^+\nu\pi^+\pi^-$	$0.9(4)\times10^{-5}$
			$e^+\nu$	$1.54(09)\times10^{-5}$
			$e^+\nu\gamma$	$1.62(47)\times10^{-5}$
			$\pi^+\pi^o\gamma$	$2.75(16)\times10^{-4}$
			$\pi^+\pi^+\pi^-\gamma$	$1.0(4)\times10^{-4}$
			$e^+\nu\pi^o\gamma$	$3.7(14)\times10^{-4}$
			$e^+e^-\pi^+$	$2.6(5)\times10^{-7}$
			$\mu^+\nu e^+e^-$	$11(3)\times10^{-7}$
			$e^+\nu e^+e^-$	$2(+2)(-1)\times10^{-7}$
K^o, $\bar{K}^o$	497.67(13) :			
K^o_S		$0.8923(22)\times10^{-10}$ sec	$\pi^+\pi^-$	68.61(24)
			$\pi^o\pi^o$	31.39(24)
			$\pi^+\pi^-\gamma$	$1.85(10)\times10^{-3}$

Table 7.II (Continued)

Particle	Mass	Mean Life or Full Width	Decay Modes	Fraction
K_L^o		$5.183(40)\times10^{-8}$ sec	$\pi^o\pi^o\pi^o$	21.5(7)
			$\pi^+\pi^-\pi^o$	12.39(18)
			$\pi^{\pm}\mu^{\pm}\nu$	27.0(5)
			$\pi^{\pm}e^{\pm}\nu$	38.8(5)
	note: $K_S^o-K_L^o$ =		$\pi e\nu\gamma$	1.3(8)
			$\pi^+\pi^-$	0.203
	$0.5349(22)\times10^{10}$ ħ sec^{-1}		$\pi^o\pi^o$	0.094(18)
			$\pi^+\pi^-\gamma$	$6(2)\times10^{-5}$
	K^o and $\bar{K}^o$ decay 50% into		$\gamma\gamma$	$4.9(5)\times10^{-4}$
	K_S^o and 50% into K_L^o		$\mu^+\mu^-$	$9.1(1.8)\times10^{-9}$
$\eta'(958)$	957.6(3)	< 1 MeV	$\eta\pi\pi$	66.2(1.7)
			$\rho^o\gamma$	29.8(1.7)
			$\omega\gamma$	2.1(4)
			$\gamma\gamma$	2.0(3)
$\rho(770)$	776	155(3) MeV	$\pi\pi$	~100%
			$\pi\gamma$	.024(7)
			e^+e^-	.0043(5)
			$\mu^+\mu^-$	.0067(12)
			$\eta\gamma$	(seen)
$\omega(783)$	782.6(3)	10.1(3) MeV	$\pi^+\pi^-\pi^o$	89.9(6)
			$\pi^+\pi^-$	1.3(3)
			$\pi^o\gamma$	8.8(5)
			e^+e^-	.0076(17)
			$\eta\gamma$	(seen)
$\Phi(1020)$	1019.6(2)	4.1(2) MeV	K^+K^-	48.6(1.2)
			$K_L^oK_S^o$	35.1(1.2)
			$\pi^+\pi^-\pi^o$	14.7
			$\eta\gamma$	1.6(2)
			$\pi^o\gamma$	0.14(5)
			e^+e^-	.031(1)
			$\mu^+\mu^-$	.025(3)
$K^*(892)$	892.2(4)	49.5(1.5) MeV	$K\pi$	~100
			$K\gamma$	.15(7)

7.2 The ω-ϕ Problem.

Several methods have been suggested for describing the low-lying mesons in twistor terms. Evidently, one requires a scheme of considerable generality, since, in addition to the low-lying states, there are many many other mesons as well. One method which has been proposed is to treat mesons as holomorphic functions of three twistors and to consider the spinor coefficient structure $\hat{\pi}^{iA}\pi_{jA'}$. This spinor coefficient structure produces a multiplet of states $\emptyset^{iA}_{jA'}$, by means of the contour integral formula

$$(7.2.1) \qquad \emptyset^{iA}_{jA'}(x) = \oint \rho_x \hat{\pi}^{Ai}\pi_{A'j} f(Z^{\alpha}_i)\Delta\pi \quad .$$

By taking the divergence of $\emptyset^{iA}_{jA'}$ we obtain a set of spin 0 mesons, and by taking the divergence-free part of $\emptyset^{iA}_{jA'}$ we obtain a set of spin 1 mesons.

Unfortunately, this procedure leads to two grave difficulties. The first problem is concerned with the spin 0^- mesons. These mesons are supposed to exhibit negative intrinsic parity. Now in the naive quark model there is no problem, because quarks have P = 1, and antiquarks have P = -1. Therefore, if they are in an S-state (i.e., no orbital angular momentum) then the combined pair will automatically have negative intrinsic parity. If one considers the spin 0^- SU(3) singlet state produced in (7.2.1), then it will be observed that what is actually being produced is the <u>derivative</u> of the field $\emptyset$ defined by

$$(7.2.2) \qquad \emptyset(x) = \oint \rho_x f(Z^{\alpha}_i)\Delta\pi \quad .$$

In other words, we have the formula

$$(7.2.3) \qquad \emptyset^{jA}_{jA'}(x) = i\nabla^{A}_{A'}\emptyset(x) \quad ,$$

which follows at once as a consequence of (7.2.1), (7.2.2), and (5.3.2). Since $\emptyset(x)$ exhibits no quark structure whatsoever in its associated contour integral formula, it is very difficult to make a case for its being of negative intrinsic parity. The second problem is concerned with the spin 1^- mesons. Here matters are even worse! According to formula (7.2.3), a spin one singlet state simply does not exist within a three-twistor framework[1]. This result corroborates Theorem (6.5.3), and, in particular, formulae (6.5.6). Thus, as matters stand we cannot account for both ω and φ.

7.3 Mesons as Quark-Antiquark Systems.

So, back to the drawing board. In Section 6.6 we discussed the fact that baryons need not be treated in any sense as bound states of quarks—at least, insofar as the low-lying baryons are concerned. With mesons the state of affairs is rather different, and indeed quite a reasonable picture of the low-lying mesons can be built up by following the quark model as closely as possible. In particular, the defects mentioned in the previous section can be eliminated.

For a single quark state, if such states exists, the relevant contour integral formula is

$$q^{Ai}(x) = \oint \rho_x \hat{\pi}^{Ai} f(Z)\Delta\pi \quad , \qquad (i = 1,2,3) \quad , \tag{7.3.1}$$

and for antiquarks the relevant formula is:

$$q_i^{A'}(x) = \oint \rho_x \pi_i^{A'} f(Z)\Delta\pi \quad . \tag{7.3.2}$$

Now in order to characterize a bound state of a quark and an antiquark we require a function of <u>six</u> twistors $f(Z^\alpha_{1i}, Z^\alpha_{2i})$, where three of the twistors refer to a quark, and the other three refer to an antiquark.

To simplify the notation in what follows, let us write

$$\begin{cases} Z^\alpha_{1i} = (\alpha^A_i \, , \alpha_{A'i}) \\ Z^\alpha_{2i} = (\beta^A_i \, , \beta_{A'i}) \end{cases} \tag{7.3.3}$$

for the spinor parts of Z^α_{1i} and Z^α_{2i} , and write

$$\begin{cases} \hat{\alpha}^{Ai} = -\partial/\partial\alpha_{Ai} \\ \hat{\beta}^{Ai} = -\partial/\partial\beta_{Ai} \end{cases} \tag{7.3.4}$$

for the associated spinor operators. Then for a quark-antiquark system we could take the spinor coefficient structure

(7.3.5) $$\hat{\alpha}^{Ai}\beta_{A'j} \quad .$$

On the other hand, we might equally well take

(7.3.6) $$\alpha_{A'j}\hat{\beta}^{Ai} \quad .$$

Which of the two do we take? Or do we, perhaps, take some linear combination? What a dilemma.

But wait! There is one quantum number which we have not yet taken into account: namely, the charge conjugation number. Charge conjugation is defined to be the operator which changes particles into antiparticles, and vice-versa, but whilst at the same time preserving handedness. Evidently we have $\hat{C}^2 = 1$; thus for eigenstates of $\hat{C}$ we must have $C = \pm 1$ for the eigenvalue. Selection rules in strong and electromagnetic processes allow one to empirically determine C for a variety of particles; for example, the photon, the ρ-meson, the ω-meson, and the φ-meson all have $C = -1$; and the pion, the η-meson, and η'-meson all have $C = 1$. Now, let us define the following operators[2]:

(7.3.7) $$\begin{cases} \hat{\alpha}^{A'i} = \kappa^{-1}\hat{P}_A^{A'}\hat{\alpha}^{Ai} \quad , & \hat{\beta}^{A'i} = \kappa^{-1}\hat{P}_A^{A'}\hat{\beta}^{Ai} \\ \alpha_{Ai} = -\kappa^{-1}\hat{P}_A^{A'}\alpha_{A'i} \quad , & \beta_{Ai} = -\kappa^{-1}\hat{P}_A^{A'}\beta_{A'i} \end{cases}$$

Under charge conjugation we have the following transformations in the spinor coefficient structure:

(7.3.8) $$\begin{cases} \hat{C}\hat{\alpha}^{Ai} = \alpha_i^A & \hat{C}\hat{\beta}^{Ai} = \beta_i^A \\ \hat{C}\alpha_i^A = \hat{\alpha}^{Ai} & \hat{C}\beta_i^A = \hat{\beta}^{Ai} \end{cases}$$

(7.3.9) $$\begin{cases} \hat{C}\hat{\alpha}^{A'i} = \alpha_i^{A'} & \hat{C}\hat{\beta}^{A'i} = \beta_i^{A'} \\ \hat{C}\alpha_i^{A'} = \hat{\alpha}^{A'i} & \hat{C}\beta_i^{A'} = \hat{\beta}^{A'i} \end{cases}$$

These formula interchange right-handed quarks with right-handed antiquarks, and interchange left-handed quarks with left-handed antiquarks, as desired. Note, however, that they do not in any way intermingle the $Z^{\alpha}_{1^i}$ twistors with the $Z^{\alpha}_{2^i}$ twistors.

Now we shall consider the spinor coefficient structure

$$\hat{\alpha}^{iA}\beta^{B}_{j} - \hat{\beta}^{iA}\alpha^{B}_{j} \quad , \tag{7.3.10}$$

where, for convenience, we have made everything left-handed. This structure is not in an eigenstate of $\hat{C}$, for when $\hat{C}$ operates on (7.3.10) we get

$$\alpha^{A}_{j}\hat{\beta}^{iB} - \beta^{A}_{j}\hat{\alpha}^{iB} \quad , \tag{7.3.11}$$

which is clearly quite distinct from (7.3.10).

However, suppose we split (7.3.10) into its spin 0 and spin 1 parts. Then we have

$$\hat{\alpha}^{iA}\beta_{jA} - \hat{\beta}^{iA}\alpha_{jA} \quad \ldots\ldots \text{ spin } 0 \quad , \tag{7.3.12}$$

and

$$\hat{\alpha}^{i(A}\beta^{B)}_{j} - \hat{\beta}^{i(A}\alpha^{B)}_{j} \quad \ldots\ldots \text{ spin } 1 \quad . \tag{7.3.13}$$

Behold! The spin 0 part is in an eigenstate of $\hat{C}$ with $C = 1$, and the spin 1 part is in an eigenstate of $\hat{C}$ with $C = -1$. These eigenvalues are indeed the observed eigenvalues. Evidently then, the minus sign taken in the superposition (7.3.10) is the correct choice.

It should be noted that the charge conjugation number selects out the spinor coefficient structure (7.3.10) quite uniquely for the low-lying mesons. If, for instance, one had chosen $\hat{\alpha}^{iA}\alpha^{B}_{j}$ or $\hat{\beta}^{iA}\beta^{B}_{j}$, or any linear combination of these two expressions, then, as a short calculation will reveal, the wrong charge conjugation numbers would have emerged.

7.4 Orbital Angular Momentum.

In order to pursue the matter of excitations of hadronic systems it is necessary as a preliminary measure to make a few remarks concerning orbital angular momentum. Suppose we consider a two-point field $\emptyset(x,y)$ satisfying the massive wave equation individually with respect to x^a and y^a, with the same mass:

$$(7.4.1)\qquad \begin{cases} (\overset{1}{\nabla}_a \overset{1}{\nabla}{}^a + m^2)\,\emptyset(x,y) = 0 \quad , & \overset{1}{\nabla}_a = \partial/\partial x^a \\ (\overset{2}{\nabla}_a \overset{2}{\nabla}{}^a + m^2)\,\emptyset(x,y) = 0 \quad , & \overset{2}{\nabla}_a = \partial/\partial y^a \quad . \end{cases}$$

Furthermore, let us suppose that $\emptyset(x,y)$ as a whole is in a definite state of total mass—that is to say, we have

$$(7.4.2)\qquad [(\overset{1}{\nabla}_a + \overset{2}{\nabla}_a)(\overset{1}{\nabla}{}^a + \overset{2}{\nabla}{}^a) + M^2]\,\emptyset(x,y) = 0 \quad ,$$

where M is the total mass. Let us denote by V the subspace $x = y$, and write ρ_V for the restriction down to V. Thus, we have, for example, $\rho_V \emptyset(x,y) = \emptyset(x,x)$. We define $\hat{L}_a$ by the relation

$$(7.4.3)\qquad \hat{L}_a = i(\overset{1}{\nabla}_a - \overset{2}{\nabla}_a) \quad .$$

Note that, on account of equations (7.4.1) and (7.4.2) we obtain:

$$(7.4.4)\qquad \hat{L}_a \hat{L}^a\,\emptyset(x,y) = (4m^2 - M^2)\,\emptyset(x,y) \quad .$$

With this information at our disposal, the following result can now be established:

7.4.5 Theorem. A two-point massive state satisfying equations (7.4.1) and (7.4.2) has a definite total spin s if and only if the expression

$$(7.4.6)\qquad \rho_V\,[\hat{L}_a \hat{L}_b \ldots \emptyset(x,y)] =: \emptyset_{ab\ldots}(x)$$

is non-vanishing if $\hat{L}_a$ occurs s times, and vanishes otherwise.

Proof. It is straightforward to verify that if $\emptyset(x,y)$ satisfies (7.4.2) then if we apply $\hat{L}_a$ to $\emptyset(x,y)$ any number of times the resulting state also satisfies equation (7.4.2). Moreover, if we assume that $\emptyset(x,y)$ is Fourier analyzable, then we have the formula

$$(7.4.7)\qquad \nabla_a \rho_V \emptyset(x,y) = \rho_V(\overset{1}{\nabla}_a + \overset{2}{\nabla}_a)\emptyset(x,y) \quad ,$$

from which, using (7.4.2), we obtain:

$$(7.4.8)\qquad (\nabla_c\nabla^c + M^2)\emptyset_{ab...}(x) = 0 \quad .$$

In order to establish that $\emptyset_{ab...}$ is in a definite spin state we must verify that it is symmetric, tracefree, and divergence-free. It is, of course, symmetric. To prove that it is tracefree one uses (7.4.4) along with the assumption that (7.4.6) vanishes if evaluated with any number of occurrences of $\hat{L}_a$ other than s. Finally, the fact that it is divergence-free follows from (7.4.7) in combination with the identity

$$(7.4.9)\qquad (\overset{1}{\nabla}{}^a + \overset{2}{\nabla}{}^a)\hat{L}_a\emptyset(x,y) = 0 \ . \quad \square$$

According to Theorem 7.4.5, the operator $\hat{L}_a$ acts as a "projection operator" for orbital angular momentum. It can also be used in our twistor contour integral formulae. For if a two-particle wave function $f(\underset{1}{Z}{}^\alpha_i, \underset{2}{Z}{}^\alpha_i)$ is in a definite state of orbital angular momentum between the two subsystems, then in evaluating the associated contour integral we must apply the operator

$$(7.4.10)\qquad \hat{L}_a = \hat{P}_{1a} - \hat{P}_{2a}$$

an appropriate number of times before applying the restriction operator ρ_x. Note, in particular, that if we denote by $\overset{1}{\rho}_x$ the operator which restricts $\underset{1}{Z}{}^\alpha_i$ down to $x^{AA'}$, and we denote by $\overset{2}{\rho}_y$ the operator which restricts $\underset{2}{Z}{}^\alpha_i$ down the $y^{AA'}$, then we have the relation

$$(7.4.11)\qquad i(\overset{1}{\nabla}_a - \overset{2}{\nabla}_a)\overset{1}{\rho}_x\overset{2}{\rho}_y f(\underset{1}{Z},\underset{2}{Z}) = \overset{1}{\rho}_x\overset{2}{\rho}_y \hat{L}_a f(\underset{1}{Z},\underset{2}{Z}) \quad ,$$

which establishes the connection between (7.4.3) and (7.4.10).

The discussion above has been limited to the case where the two masses of the subsystems are identical. It is not especially difficult to generalize all the relevant formulae so that they apply when the two masses are distinct. In the sections which follow $\hat{L}_a$ is to be understood as being defined in such a way as to

Table 7.III: Mesons and Meson Resonances

Non-strange Mesons

State	$I^G(J^P)C_n$	State	$I^G(J^P)C_n$
π	$1^-(0^-)+$	$\omega(1670)$	$0^-(3^-)^-$
η	$0^+(0^-)^+$	$g(1680)$	$1^+(3^-)^-$
$\rho(770)$	$1^+(1^-)^-$	$\rightarrow\chi(1690)$	
$\omega(783)$	$0^-(1^-)^-$	$\rightarrow A_4(1900)$	1^-
$\rightarrow$M(940-953)		$\rightarrow$X(1900)	$1^-(4^+)^+$
$\eta'(958)$	$0^+(0^-)^+$	S(1935)	1
$\varepsilon(980)$	$1^-(0^+)^+$	h(2040)	$0^+(4^+)^+$
S*(980)	$0^+(0^+)^+$	T(2190)	$1^+(3^-)^-$
$\rightarrow$H(990)		U(2350)	$0^+(4^+)^+$
$\phi(1020)$	$0^-(1^-)^-$	$\rightarrow N\bar{N}(2360)$	1
$\rightarrow$M(1033)-1040)		$\rightarrow N\bar{N}(1400\text{-}3600)$	
$\rightarrow\eta_N(1080)$	$0^+(N)^+$	$\rightarrow$X(1900-3600)	
$A_1(1100)$	$1^-(1^+)^+$	$\rightarrow e^+e^-(1100\text{-}3100)$	
$\rightarrow$M(1150-1170)		$\rightarrow\chi(2830)$	
B(1235)	$1^+(1^+)^-$	$\psi(3100)$	$0^-(1^-)^-$
$\rightarrow\rho'(1250)$	$1^+(1^-)^-$	$\chi(3415)$	$0^+(0^+)^+$
f(1270)	$0^+(2^+)^+$	$\rightarrow\chi(3455)$	
D(1285)	$0^+(A)^+$	$\chi(3510)$	$0^+(A)^+$
$\varepsilon(1300)$	$0^+(0^+)^+$	$\chi(3555)$	$0^+(N)^+$
$A_2(1310)$	$1^-(2^+)^+$	$\chi(3685)$	$0^-(1^-)^-$
E(1420)	$0^+(A)^+$	$\psi(3770)$	$(1^-)^-$
$\rightarrow\chi(1410\text{-}1440)$		$\rightarrow\psi(4030)$	$(1^-)^-$
f'(1515)	$0^+(2^+)^+$	$\psi(4415)$	$(1^-)^-$
$\rightarrow F_1(1540)$	1(A)	$\Upsilon(9500)$	$(1^-)^-$
$\rho'(1600)$	$1^+(1^-)^-$	$\Upsilon(10060)$	$(1^-)^-$
$A_3(1640)$	$1^-(2^-)^+$		

Table 7.III (Continued)

Strange Mesons

State	$I(J^P)$
K	$1/2(0^-)$
K*(892)	$1/2(1^-)$
Q_1(1280)	$1/2(1^+)$
→Q_2(1400)	$1/2(1^+)$
→K'(1400)	$1/2(0^-)$
k(1400)	$1/2(0^+)$
K*(1430)	$1/2(2^+)$
→K_N(1700)	1/2
L(1770)	1/2(A)
K*(1780)	$1/2(3^-)$
→K*(2200)	
→I(2600)	

Charmed Mesons

State	$I(J^P)$
D(1870)	$1/2(0^-)$
D*(2010)	$1/2(1^-)$
→F(2030)	
→F*(2140)	

An arrow (→) denotes a state which is not to be regarded as yet well-established.

Notation:

I = isospin

G = G-parity

J = spin

P = intrinsic parity

C_n = charge conjugation parity

N = "normal" parity (0^+ , 1^- , 2^+ , 3^- ...)

A = "abnormal" parity (0^- , 1^+ , 2^- , 3^+ ...)

take into account this generalization, when necessary.

7.5 Excited Meson States.

In this section we shall investigate the possibilities of describing "excited" meson states in twistor terms. Many meson resonances are known, and in Table 7.III the reader will find a list of the currently observed states (Particle Data Group, 1978). One approach to understanding meson excitations is to treat these states as quark-antiquark pairs with units of orbital angular momentum between the quark and the antiquark. Certain features of the resulting spectrum are not sensitive in any significant way to the specific nature of the "binding forces" between the quark and the antiquark, and it is upon these features that we shall concentrate for the moment. We cannot, for example, say anything about "radial excitations", since these depend rather critically on the type of binding that is involved.

$(J^P)C_n$	SU(3),L	I = 1	I = 0	I = 1/2
$(0^-)^+$	$\{\underline{9}^1,0^+\}$	π	η,η'	K
$(1^-)^-$	$\{\underline{9}^3,0^+\}$	$\rho(770)$	$\omega(783),\phi(1020)$	$K^*(892)$
$(1^+)^-$	$\{\underline{9}^1,1^-\}$	B(1235)		
$(0^+)^+$	$\{9^3,1^-\}$	$\delta(980)$	$S^*(980),\varepsilon(1300)$	$\kappa(1400)$
$(1^+)^+$	$\{9^3,1^-\}$	$A_1(1100)$	D(1285),E(1420)	Q
$(2^+)^+$	$\{9^3,1^-\}$	$A_2(1310)$	f(1270),f'(1515)	$K^*(1430)$

Table 7.IV

The Observed Meson Nonets

In Table 7.IV we list several of the observed meson "nonets". Quotation marks because in some cases the evidence is a bit shaky. Only one well-established state in the $(1^+)^-$ multiplet. In the $(1^+)^+$ multiplet the spin-parity of E(1420) is not known definitely—although it is known to be "abnormal" (i.e., 0^-, 1^+, 2^-, ..., etc.). There are various other problems, as well. Nevertheless, if taken at face value the picture presented in Table 7.IV is quite consistent with the notion that several of the observed excited meson states are obtained by adding in one unit of orbital angular momentum to the bound quark-antiquark system.

Let us consider first the C = +1 states. The orbital angular momentum projection operator that must appear in the spinor coefficient structure is

$$\hat{L}^{AA'} = \hat{\alpha}^{Ai}\alpha_i^{A'} - \hat{\beta}^{Ai}\beta_i^{A'} \quad , \tag{7.5.1}$$

in accordance with formula (7.4.10). In what follows it is convenient to work with the completely "left-handed" operator

$$L^{AB} = \hat{L}^{AA'}\hat{P}^{B}_{A'} \quad . \tag{7.5.2}$$

From equation (7.4.9) it follows that L^{AB} is symmetric on the indices A and B. Now it is an elementary exercise to verify that both $\hat{L}^{AA'}$ and L^{AB} are invariant under the charge conjugation transformations (7.3.8) and (7.3.9). Therefore, the spinor coefficient structure

$$[\hat{\alpha}^{i(A}\beta_j^{B)} + \hat{\beta}^{i(A}\alpha_j^{B)}]L^{CD} \tag{7.5.3}$$

is in an eigenstate of C = 1. Note that the choice of sign in (7.5.3) is opposite to that of (7.3.10). Let us introduce the convenient abbreviation

$$T_j^{iAB} = \hat{\alpha}^{i(A}\beta_j^{B)} + \hat{\beta}^{i(A}\alpha_j^{B)} \quad . \tag{7.5.4}$$

Then expression (7.5.3) can be reduced to three distinct spin states, as follows:

$$T_j^{iAB}L_{AB} \quad \ldots\ldots\ldots\ldots\ldots\ldots \quad (0^+)^+ \tag{7.5.5}$$

$$T_j^{iC(A}L^{B)}_{C} \quad \ldots\ldots\ldots\ldots\ldots\ldots \quad (1^+)^+ \tag{7.5.6}$$

(7.5.7) $$T_j^{i(AB_L CD)} \ldots\ldots\ldots\ldots\ldots\ldots (2^+)^+ \quad .$$

Now let us consider the C = -1 case. If we define a quantity T by

(7.5.8) $$T = \hat{\alpha}^{iA}\beta_{jA} + \hat{\beta}^{iA}\alpha_{jA} \quad ,$$

then the spinor coefficient structure that we desire is given by

(7.5.9) $$TL^{AB} \ldots\ldots\ldots\ldots\ldots\ldots\ldots\ldots (1^+)^- \quad .$$

Here, of course, we use the fact that $\hat{C}T = -T$, as can be readily verified.

All four of the multiplets (7.5.5), (7.5.6), (7.5.7), and (7.5.9) have positive intrinsic parity. This is obtained as a product of the inherent negative intrinsic parity of the quark-antiquark system, and the negative parity of the single unit of orbital angular momentum.

A very curious feature of the preceeding material is illustrated in the fact that all six of the multiplets (7.3.12), (7.3.13), (7.5.5)-(7.5.7), and (7.5.9) have the property of being antisymmetric under the interchange of the labels α and β. This would appear to be an outcrop of Fermi statistics. Note that at no stage have we actually imposed Fermi statistics—our spinor coefficient structures have been designed purely on the basis of phenomenological considerations. Nevertheless it does seem reasonable hereon out to insist—especially in the cases where the empirical data is scantly or ambiguous—that our spinor coefficient structures (and a fortiori, the associated twistor functions) exhibit appropriate statistical properties.

In connection with charmed particles and the ψ/J family one can pursue the matter of meson excitations one step further. For a single charmed quark it is necessary, apparently, to consider a function of four twistors transforming under the action of the group SU(4). Appropriate meson states can be built up as quark-antiquark pairs describable in terms of functions of eight twistors. This topic will be discussed elsewhere.

7.6 Baryon Resonances.

In the case of mesons and meson excitations our twistor model does not differ all that much from the standard quark model. The principal significant difference lies in the lack of internal color degrees of freedom. These hypothetical color degrees of freedom, which would certainly be of relevance for dynamical processes, can be treated as insignificant insofar as certain general features of the meson spectrum are concerned—to that extent, therefore, we have an approximate agreement between twistor theory and the "old" theory.

In the case of baryons and baryon excitations, however, our model does indeed differ in several substantial ways from the picture suggested by the standard naive colored quark model. Let me describe in loose physical terms the set-up that I envisage before getting involved in details of a more technical character:

As far as hadrons are concerned we hypothesize that there are, in nature, three basic "primitive" particle types. These are called quarks, diquarks, and triquarks. In addition, we have the corresponding antiparticles. A diquark is not in any sense to be regarded as a bound state of two quarks, nor is a triquark to be regarded as a bound state of three quarks. Diquarks and triquarks are particle types quite distinct from quarks. Insofar as their internal constitution is concerned, quarks exhibit more or less the same degree of complexity as do diquarks and triquarks. The low-lying baryons are examples of triquarks. Many of the observed mesons are quark-antiquark bound states. It is not out of the question that some mesons are formed as diquark-antidiquark bound states, or possibly as other "exotic" combinations. We propose that many of the observed baryon resonances are quark-diquark bound states. These resonances can be formed, for example, as follows. We collide, say, a baryon and a meson. The antiquark component of the meson system interacts with the baryon triquark so as to produce a diquark. This leaves us, then, with a quark-diquark system. The quark-diquark bound state is unstable, and as soon as the vacuum can produce a quark-antiquark pair the whole process reverses, and the resonant state disintegrates. Remarkably enough, this simple picture can account for a wealth of data.

In Table 7.V we summarize the known baryon excitation states, as catalogued in the Review of Particle Properties[4]. The reader cannot help but be amazed when confronted with this vast list. Here more than anywhere we have evidence for the rich internal structure of elementary particles. Many an hour can be spent musing over the intricacies of this table, where many a symmetry lies submerged and half-hidden, like precious shells half-hidden in the sand on a beach. Take care! What one moment we think is a flawless conch on closer inspection often proves to be but a piece of driftwood.

Let us now examine what sort of multiplets are obtained in the case of quark-diquark bound states. Following the pattern of Section 7.3 we shall regard these states as functions of a pair of twistor triplets, using the notation (7.3.3) and (7.3.4). Thus for a quark-diquark system we must consider spinor coefficient structures of the form

$$\hat{\alpha}^{Ai}\hat{\beta}^{Bj}\hat{\beta}^{Ck} , \tag{7.6.1}$$

where $\hat{\alpha}^{Ai}$ refers to the quark, and $\hat{\beta}^{Bj}\hat{\beta}^{Ck}$ refers to the diquark. Now strictly speaking (7.6.1) is not correct, since we have not yet taken into account the proper statistical relations that should hold between the quark and the diquark. Since the quark is a fermion and the diquark is a boson, the spinor coefficient structure (as a whole) must be <u>symmetric</u> under the interchange of the labels α and β. Therefore, we replace (7.6.1) with the expression

$$T_{\pm}^{AiBjCk} = \hat{\alpha}^{Ai}\hat{\beta}^{Bj}\hat{\beta}^{Ck} \pm \hat{\beta}^{Ai}\hat{\alpha}^{Bj}\hat{\alpha}^{Ck} , \tag{7.6.2}$$

where the plus sign is used for even orbital angular momentum, and the minus sign is used for odd orbital angular momentum.

We shall find it convenient on occasion to introduce the index clumping convention Ai = a, Bi = b, etc., and write

$$T_{\pm}^{abc} = T_{\pm}^{AiBjCk} . \tag{7.6.3}$$

The indices a, b, c, etc., are often, by abuse of terminology, called SU(6) indices.

Table 7.V: The Baryon Resonance Spectrum

Even Parity		Odd Parity		Undetermined Parity	
N(939)P11	****	N(1520)D13	****	N(3030)	***
N(1470)P11	****	N(1535)S11	****	N(3245)	*
N(1540)P13	*	N(1670)D15	****	N(3690)	*
N(1688)F15	****	N(1700)S11	****	N(3755)	*
N(1780)P11	***	N(1700)D13	***		
N(1810)P13	***	N(2040)D13	**		
N(1990)F17	**	N(2100)S11	*		
N(2000)F15	**	N(2100)D15	**		
N(2220)H19	***	N(2190)G17	***		
		N(2200)G19	***		
		N(2650)I1 11	***		
Δ(1232)P33	****	Δ(1650)S31	****		
Δ(1550)P31	*	Δ(1670)D33	***	Δ(2850)	***
Δ(1690)P33	***	Δ(1900)S31	*	Δ(3230)	***
Δ(1890)F35	****	Δ(1960)D35	**		
Δ(1910)P31	****	Δ(2160)	***		
Δ(1950)F37	****				
Δ(2420)H3 11	***				
Λ(1115)P01	****	Λ(1405)S01	****	Λ(2010)	**
Λ(1600)P01	**	Λ(1520)D03	****	Λ(2350)	****
Λ(1800)P01	**	Λ(1670)S01	****	Λ(2585)	***
Λ(1815)F05	****	Λ(1690)D03	****		
Λ(1860)P03	***	Λ(1800)G09	*		
Λ(2020)F07	*	Λ(1830)D05	****		
Λ(2110)F05	***	Λ(1870)S01	***		
		Λ(2100)G07	****		
		Λ(2325)D03	*		
Σ(1193)P11	****	Σ(1580)D13	**	Σ(1480)	*
Σ(1385)P13	****	Σ(1620)S11	**	Σ(1670)	**
Σ(1660)P11	***	Σ(1670)D13	**	Σ(1690)	**
Σ(1770)P11	*	Σ(1750)S11	***	Σ(2250)	****
Σ(1840)P13	*	Σ(1765)D15	****	Σ(2455)	***
Σ(1880)P11	**	Σ(1940)D13	***	Σ(2620)	***
Σ(1915)F15	****	Σ(2000)S11	*	Σ(3000)	**
Σ(2030)F17	****	Σ(2100)G17	*		
Σ(2070)F15	*				
Σ(2080)P13	**				
Ξ(1317)P11	****			Ξ(1630)	**
Ξ(1530)P13	****			Ξ(1820)?13	***
				Ξ(1940)	**
				Ξ(2030)?1?	***
				Ξ(2120)	*
Ω(1672)P03	****			Ξ(2250)	*
				Ξ(2500)	**

Note that we have the relation $T_{\pm}^{abc} = T_{\pm}^{acb}$. Accordingly, we see that T^{abc} can be split into precisely two distinct parts, each of which exhibits definite Young tableau symmetry, as follows:

(7.6.4) $$F_{\pm}^{abc} = T_{\pm}^{(abc)} \quad ,$$

(7.6.5) $$G_{\pm}^{abc} = T_{\pm}^{[ab]c} \quad .$$

A straightforward calculation shows that $F_{\pm}^{abc}$ has exactly 56 independent components, and that $G_{\pm}^{abc}$ has exactly 70 independent components.

We see that our basic quark-diquark system splits into a $\underline{56}$ and a $\underline{70}$. To these supermultiplets we can then begin to add units of orbital angular momentum, and in this way produce the following set of supermultiplets:

(7.6.6) $$\{\underline{56},0^+\}' \ , \ \{\underline{56},1^-\} \ , \ \{\underline{56},2^+\} \ , \ \{\underline{56},3^-\} \ , \ \ldots$$

(7.6.7) $$\{\underline{70},0^+\} \ , \ \{\underline{70},1^-\} \ , \ \{\underline{70},2^+\} \ , \ \{70,3^-\} \ , \ \ldots \quad .$$

We have put a prime, incidentally, on the supermultiplet $\{\underline{56},0^+\}'$, so as to distinguish it clearly from the more basic $\{\underline{56},0^+\}$ triquark supermultiplet (spinor coefficient structure $\hat{\pi}^{Ai}\hat{\pi}^{Bj}\hat{\pi}^{Ck}$) to which the low-lying baryons belong.

J^P , SU(3)	N,Δ	Σ	Λ
$1/2^-$ $\underline{8}^2,\underline{1}^2$	N(1535)S11	Σ(1620)S11	Λ(1405)S01, Λ(1670)S01
$1/2^-$ $\underline{8}^4$	N(1700)S11		Λ(1870)S01
$3/2^-$ $\underline{8}^2,\underline{1}^2$	N(1520)D13	Σ(1580)D13	Λ(1690)D03, Λ(1520)D03
$3/2^-$ $\underline{8}^4$	N(1700)D13	Σ(1670)D13	
$5/2^-$ $\underline{8}^4$	N(1670)D15	Σ(1765)D15	Λ(1830)D05
$1/2^-$ $\underline{10}^2$	Δ(1650)S31	Σ(1750)S11	
$3/2^-$ $\underline{10}^2$	Δ(1670)D33		

Table 7.VI

The Observed $\{70,1^-\}$ Baryon Supermultiplet

The evidence for the existence of a $\{\underline{56},0^+\}'$ is quite good, including as nucleonic members the N(1470)P11 and the Δ(1690)P33 (cf. below). Evidence for the $\{\underline{70},0^+\}$ and $\{\underline{56},1^-\}$ is very tenuous, although it is not altogether implausible that these states should exist. The $\{\underline{70},1^-\}$ is almost completely well-established now, and we list the observed members of this supermultiplet in Table 7.VI. The $\{\underline{56},2^+\}$ is reasonably well-established, with the following nucleonic content:

(7.6.8)

$\underline{10}^4$ $7/2^+$ .... Δ(1950)F37

$\underline{10}^4$ $5/2^+$ .... Δ(1890)F35

$\underline{10}^4$ $3/2^+$ .... Δ(1690)P33 (?)

$\underline{10}^4$ $1/2^+$ .... Δ(1910)P31

(7.6.9)

$\underline{8}^2$ $5/2^+$ .... N(1688)F15

$\underline{8}^2$ $3/2^+$ .... N(1810)P13

It is possible that the Δ(1690)P33 may include more than one resonance in or near the 1650-1900 MeV mass range, and in the assignments above it is presumed that one of these resonances belongs to the $\{\underline{56},0^+\}$, and another belongs to the $\{\underline{56},2^+\}$.

The evidence for higher supermultiplets is sparse and relatively ambiguous, but there is no doubt that a number of such supermultiplets do, in fact, exist. As for the $\{\underline{70},0^+\}$ and the $\{\underline{56},1^-\}$, we can only speculate. The model does indeed seem to imply the existence of these states, although one might envisage the possibility of some modification which would eliminate them.

In addition to the supermultiplets listed in (7.6.6) and (7.6.7) one might expect further states on the basis of radial excitations. As explained earlier, it is beyond our scope at present to make any definitive assertions on that score. These matters, it is hoped, will be pursued elsewhere.

7.7 The Deuteron.

To conclude this chapter, we shall discuss briefly certain aspects of the representation of the deuteron (bound state of a proton and a neutron) in twistor terms. The properties of this particle are of considerable interest in a number of areas outside of elementary particle physics proper (e.g., astrophysics, nuclear physics, etc.) and thus it seems worthwhile to point out here that there is some scope for effectively handling the deuteron (and possibly other light nuclei) within the twistor framework.

The proton has the spinor coefficient structure $u^A u^B d_B$, and the neutron has the spinor coefficient structure $d^A d^B u_B$. In order to consider a bound state of the two we must take a holomorphic function of six twistors $f(\underset{1}{Z}{}^\alpha_i, \underset{2}{Z}{}^\alpha_i)$ and consider the spinor coefficient structure

$$(7.7.1) \qquad T^{AB}_{\pm} := \underset{1}{u}{}^C \underset{1}{d}{}_C \underset{1}{u}{}^{(A} \underset{2}{d}{}^{B)} \underset{2}{d}{}^D \underset{2}{u}{}_D \mp \underset{2}{u}{}^C \underset{2}{d}{}_C \underset{2}{u}{}^{(A} \underset{1}{d}{}^{B)} \underset{1}{d}{}^D \underset{1}{u}{}_D \quad .$$

Since the individual nucleons are fermions one uses T^{AB}_{+} in conjunction with even units of orbital angular momentum, and T^{AB}_{-} in conjunction with odd units of orbital angular momentum.

The deuteron is known to be predominantly S-state, i.e., zero orbital angular momentum; thus T^{AB}_{+} is the primary contribution to the complete spinor coefficient structure. There is, however, a small admixture of D-state into the ground state wave function of the deuteron. This can be inferred from its magnetic moment and its electric quadrupole moment. A D-state corresponds to the spinor coefficient structure

$$(7.7.2) \qquad T^{CD}_{+} L_C{}^A L_D{}^B \quad ,$$

where L_{AB} is the orbital angular momentum projection operator. Note that the spinor indices in (7.7.2) have been contracted so as to produce a state of spin 1. The complete spinor coefficient structure of the deuteron is then of the form

$$(7.7.3) \qquad \xi T^{AB}_{+} + \eta T^{CD}_{+} L_C{}^A L_D{}^B \quad ,$$

where ξ and η are appropriate numerical coefficients, weighting the two contributions.

It is possible, in accordance with the general principles layed out in the previous chapters, to construct a set of holomorphic differential operators which act as observables for the deuteron. We require that $f(Z^{\alpha}_{1i}, Z^{\alpha}_{2i})$ be placed in an eigenstate compatible with the spinor coefficient structure (7.7.3).

Chapter 7, Notes

1. The ω-ϕ problem was first pointed out by G.A.J. Sparling in early 1975.

2. Much of the material in this section, as well as the next four sections, was formulated in collaboration with M. Sheppard. I would also like to express my gratitude to A. Popovich, who stressed the importance of charge conjugation to me, for numerous very useful discussions and suggestions in connection with the material described in this chapter.

3. The constant κ is $m/\sqrt{2}$. The signs in these formulae have been chosen such that the momentum operator $\hat{P}_{AA'}$ is invariant under transformations (7.3.8) and (7.3.9).

4. Notation: N(1688)F15

- 5 — twice the spin
- 1 — isospin (times 2 for N, Δ, and Ξ)
- F — partial wave
- 1688 — nominal mass
- N — particle type (N, Δ, Λ, Σ, Ξ, or Ω) .

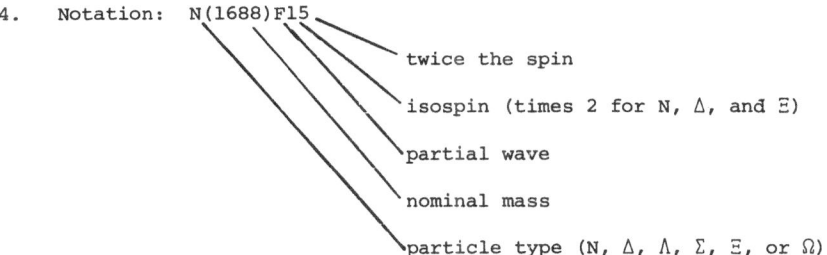

CHAPTER 8

LEPTONS AND WEAK INTERACTIONS

8.1 Properties of Leptons.

There are eight well-known "old" lepton states. The names and symbols for these states are as follows:

(8.1.1)

e^- ..... the electron	e^+ ..... the positron
μ^- ..... the muon	μ^+ ..... the antimuon
ν_e ..... the electron neutrino	$\bar{\nu}_e$ ..... the electron antineutrino
ν_μ ..... the muon neutrino	$\bar{\nu}_\mu$ ..... the muon antineutrino .

In 1976 a new lepton state called the τ-particle was discovered. It may also have a neutrino associated with it. There may be more leptons yet to be discovered.

Perhaps the best understood of the four neutrino states is the electron antineutrino $\bar{\nu}_e$. This massless (or at least nearly massless) particle is emitted as a companion to the electron in ordinary β^- radioactive decay. For example, the hydrogen isotope tritium ${}_1H^3$ decays according to the scheme

$$ {}_1H^3 \longrightarrow {}_2He^3 + e^- + \bar{\nu}_e \quad , \tag{8.1.2} $$

the products of the decay being an electron, an electron antineutrino, and a helium isotope. Another example is the decay of the ${}_6C^{14}$ isotope of carbon:

$$ {}_6C^{14} \longrightarrow {}_7N^{14} + e^- + \bar{\nu}_e \quad . \tag{8.1.3} $$

And indeed, even the neutron itself undergoes β^- decay:

$$ N \longrightarrow P + e^- + \bar{\nu}_e \quad . \tag{8.1.4} $$

Many examples of β^- decay are known and have been studied; almost all neutron-rich isotopes can undergo β^- decay. The detection of the electron antineutrino was first achieved by Reines and Cowan (1953). A nuclear reactor was employed for the production of a suitable flux of electron antineutrinos, and they searched for the following reaction:

(8.1.5) $$\bar{\nu}_e + P \longrightarrow e^+ + N \quad .$$

In spite of the very low cross-section for this interaction the experiment emerged a success on account of the elegant techniques which were developed for the detection of simultaneously produced neutrons and positrons.

The electron neutrino ν_e is emitted in the β^+ decay of proton-rich nuclei. In these decays one of the nuclear protons is converted into a neutron, with the emission of a positron and an electron neutrino. The carbon isotope ${}_6C^{11}$ decays by β^+ emission, for example, as follows:

(8.1.6) $${}_6C^{11} \longrightarrow {}_5B^{11} + e^+ + \nu_e \quad .$$

A process closely related to β^+ decay is K-capture. In this case, instead of emitting a positron the nucleus grabs an electron from the lowest atomic electron shell (the K-shell). One of the nuclear protons is converted into a neutron, and an electron neutrino is emitted. An example of K-capture is found in the atom of the europium isotope ${}_{63}Eu^{152}$, which decays according to the following scheme:

(8.1.7) $$e^- + {}_{63}Eu^{152} \longrightarrow \nu_e + {}_{62}\overset{*}{Sm}{}^{152}(1^-)$$
$$\hookrightarrow {}_{62}Sm^{152}(0^+) + \gamma \quad .$$

The K-capture results initially in a spin-parity 1^- excited state of the Sm^{152} nucleus; this nucleus then shifts its configuration to that of the 0^+ ground state, with the emission of a photon (γ) of a characteristic energy. This rather exotic europium decay was the decay analysed in the remarkable experiment of Goldhaber, Grodzins, and Sunyar (1958). This experiment led to the very curious conclusion that the <u>electron neutrino is inherently in a state of negative helicity</u>, i.e., spins in a left-handed fashion. This fact is a manifestation of the breakdown of space-reflection symmetry in weak interactions.

Another example of K-capture decay occurs in the case of the radioactive isotope ${}_{18}Ar^{37}$ of argon:

(8.1.8) $$e^- + {}_{18}Ar^{37} \longrightarrow e^o + {}_{17}Cl^{37} \quad .$$

The inverse of this reaction can, in principle, be used to detect electron neutrinos. The famous experiments of R. Davis and coworkers were originally designed to show that ν_e and $\bar{\nu}_e$ were indeed distinct types of particles. It was known that large fluxes of antineutrinos are emitted from nuclear reactors, and the question was whether those antineutrinos would initiate the inverse of the reaction cited above. The negative conclusion of this experiment showed that ν_e is, at least in some sense, a distinct particle state from $\bar{\nu}_e$. A variant on this experiment has led to the celebrated solar-neutrino problem. The reaction

(8.1.9) $$e^o + {}_{17}C^{37} \longrightarrow e^- + {}_{18}Ar^{37}$$

can be used to detect electron neutrinos emitted from the sun. Although the reaction (8.1.9) is indeed observed, it is not entirely clear that it is sufficiently often observed to account satisfactorily for the neutrino flux anticipated on the basis of standard models of stellar structure.

The muon neutrino and antineutrino are produced in the decay products of numerous elementary particles, perhaps most notably in the decays of the charged pions:

(8.1.10) $$\pi^+ \longrightarrow \mu^+\nu_\mu \quad , \quad \pi^- \longrightarrow \mu^-\bar{\nu}_\mu \quad ,$$

these being the principal decay modes of the charged pions. These decay modes played a role in the experiment of Danby et al (1962) which showed that muon neutrinos are in some sense distinct both from ν_e and $\bar{\nu}_e$. The neutrinos produced in the decay-in-flight of π^+ particles were used to bombard nuclei, and a search was made for the following hypothetical reactions:

(8.1.11) $$\nu_\mu + (Z,A) \longrightarrow (Z\mp1,A) + e^\pm \quad .$$

Neither reaction was observed. A related experiment by Borer et al (1969) indicates that ν_μ is distinct from $\bar{\nu}_\mu$. They showed that the reaction

(8.1.12) $$\nu_\mu + (Z,A) \longrightarrow (Z+1,A) + \mu^-$$

<u>does</u> take place, whereas the reaction

(8.1.13) $$\nu_\mu + (Z,A) \longrightarrow (Z-1,A) + \mu^+ \qquad (?)$$

<u>does not</u> appear to take place. Muon neutrinos are also, presumably, produced in muon β-decay:

(8.1.14) $$\mu^- \longrightarrow \nu_\mu e^- \bar{\nu}_e \quad , \quad \mu^+ \longrightarrow \bar{\nu}_\mu e^+ \nu_e \quad .$$

Unfortunately there is no <u>direct</u> evidence that the two neutrinos emitted in muon decay are specifically of the ν_μ and $\bar{\nu}_e$ sort; this is pure hypothesis. If, however, there is not in any sense a separately conserved muonic quantum number, it is difficult to imagine why the reaction $\mu^- \longrightarrow e^-\gamma$ is never observed.

It is commonly assumed that all four neutrino states are massless. This means that free neutrino states ought to be describable in terms of positive frequency solutions to the Weyl neutrino equation. The electron neutrino and the muon neutrino are both in <u>negative</u> helicity states, and thus are represented by positive frequency solutions of the ZRM equation $\nabla^{A'A}\phi_A = 0$; while the electron antineutrino and the muon antineutrino are both in <u>positive</u> helicity states, and thus are represented by positive frequency solutions of the equation $\nabla^{A'A}\phi_{A'} = 0$. It should be noted that while experimental evidence shows that neutrino and antineutrino masses must be <u>low</u>, nevertheless there is comparatively little evidence whatsoever that the neutrino and antineutrino masses are actually <u>zero</u>. The best current upper bound on the muon neutrino mass, for example, is given according to the analysis of Clark et al (1974); they conclude that the μ^o mass is 0.65 MeV, or less. Not a very stringent bound, considering that the mass of the <u>electron</u> is 0.5110034 ± .0000014 MeV.! Indeed, it would be edifying to have better experimental information on the muon neutrino mass. The same authors conclude that the $\mu^o - \bar{\mu}^o$ mass difference is .45 MeV or less. (It is quite important to have an independent determination of the $\bar{\mu}^o$ mass, since there is certainly no a priori argument that ensures the $\mu^o - \bar{\mu}^o$ mass difference will vanish!) The case for the electron antineutrino is a bit better: the upper bound on its mass, as determined by Bergkvis (1972) in his observations of tritium decay, and supported by a number of other experiments, is 0.00006 MeV. The experimental bound on the electron neutrino mass, as determined by Beck and Daniel (1968) analyzing Na^{22} decay, is 0.0041 MeV; again not so good, considering the upper bound of approximately 10^{-21} MeV currently available for the photon mass by a variety of experiments. In summary: to say that the photon has zero rest mass is one thing, to suggest that the neutrinos all have zero rest mass is another!

The tau particle was first observed by Perl et al (1975, 1976) in the debris of electron-positron annihilation. Reactions of the form

$$e^+e^- \longrightarrow \mu^{\pm}e^{\mp} + \text{(neutrals)} \tag{8.1.15}$$

were observed, and, after various alternatives had been systematically ruled out, it was deduced that the reactions took place via the following mechanisms:

(8.1.16) $e^+e^- \longrightarrow \tau^+\tau^-$

$\tau^- \longrightarrow e^-\bar{\nu}_e\nu_\tau$ or $\mu^-\bar{\nu}_\mu\nu_\tau$

$\tau^+ \longrightarrow \mu^+\nu_\mu\bar{\nu}_\tau$ or $e^+\nu_e\bar{\nu}_\tau$.

It is simply by hypothesis that the neutrinos emitted in these reactions are specifically of the type mentioned above——in particular, the evidence for a distinct tau-neutrino ν_τ is quite indirect. It does seem reasonable, however, that τ has its own conserved quantum number——otherwise, one of the decays $\tau^- \longrightarrow \mu^-\gamma$ and $\tau^- \longrightarrow e^-\gamma$ would certainly have been observed: but neither of these decays appears to occur. The τ-particle has mass 1807(20) MeV, and spin 1/2.

If we assume that the τ-particle is a "sequential" lepton——that is to say, it has its own conserved quantum number and an associated neutrino (which may or may not be massless)——then it is consistent with all the known data on leptonic and semileptonic processes to assign each lepton a set of five quantum numbers, as shown in Table 8.I:

	particles						antiparticles					
	e^-	ν_e	μ^-	ν_μ	τ^-	ν_τ	e^+	$\bar{\nu}_e$	μ^+	$\bar{\nu}_\mu$	τ^+	$\bar{\nu}_\tau$
electric charge	-1	0	-1	0	-1	0	1	0	1	0	1	0
electron number	1	1	0	0	0	0	-1	-1	0	0	0	0
muon number	0	0	1	1	0	0	0	0	-1	-1	0	0
tau number	0	0	0	0	1	1	0	0	0	0	-1	-1
lepton number	1	1	1	1	1	1	-1	-1	-1	-1	-1	-1

Table 8.I

Lepton Quantum Numbers

The fifth of these numbers—namely, lepton number—is redundant, and is given by the sum of the electron number, the muon number, and the tau number. There is nothing particularly sacred about the assignment of quantum numbers given in Table 8.I, and one might fruitfully envisage the possibility of alternative schemes.

8.2 Space Reflection Symmetry Violation.

The violation of space reflection symmetry, or "parity symmetry", as it is sometimes called, is one of the characteristic features of weak processes involving leptons and, in fact, weak processes in general. Just what is this feature, and how does it manifest itself in weak interactions? The most glaring example of parity violation occurs in connection with the neutrinos. As we have said, the neutrinos ν_e and ν_μ are always observed to be in eigenstates of negative helicity, and on the other hand the antineutrinos $\bar{\nu}_e$ and $\bar{\nu}_\mu$ are always observed to be in eigenstates of positive helicity. Parity transformations, as we shall see explicitly, carry positive helicity fields into negative helicity fields, and vice-versa. And thus the parity transformation operation is not a valid symmetry operation, at least as far as neutrino states are concerned.

A space-reflection is a transformation $x^a \longrightarrow \tilde{x}^a$ with $\tilde{x}^{AA'} = t^A_{B'} t^{A'}_B x^{BB'}$, where t^a is an arbitrary timelike vector (orthogonal to the family of three-spaces where the reflections are made) normalized $t^a t_a = 2$. If $\phi_{A'}$ is a positive helicity solution of the ZRM equation $\nabla^{AA'}\phi_{A'} = 0$, then a space-reflection transformation on $\phi_{A'}$ is given by[1]

(8.2.1) $$\phi_{A'}(x) \longrightarrow \tilde{\phi}_A(x) \quad ; \quad \tilde{\phi}_A(x) = t^{B'}_A \phi_{B'}(\tilde{x})$$

with $\tilde{x}$ given as above. It is a straightforward matter to verify that $\tilde{\phi}_A(x)$ defines a solution of the negative helicity ZRM equation. Thus, space-reflection transformations carry positive helicity fields into negative helicity fields. A more general and, in certain respects, more subtle manifestation of parity violation in weak processes is in the nature of the coupling which is responsible phenomenologically for the weak interactions. [This is the so-called universal Fermi coupling[2].] The essential point here is that while the massive free-particle Dirac equation is, in a certain sense to be described, invariant under space-reflection transformations, when those couplings are included which are responsible for weak interactions the resulting system is not invariant. When the couplings are to neutrinos it is, from what was said earlier, a foregone conclusion that invariance will be lacking, but the point here is that even when neutrinos are not involved (e.g., in hyperon decays, such as $\Lambda^0 \longrightarrow P\pi^-$) the invariance is lacking. Putting it briefly, this comes about as follows. The Dirac equation, in two-component spinor form, consists of a pair of spinors ξ_A and $\eta_{A'}$ which satisfy

(8.2.2) $$\nabla^A_{A'}\xi_A = \kappa\eta_{A'} \quad , \quad \nabla^{A'}_A \eta_{A'} = \kappa\xi_A \quad ,$$

where $\kappa = m/\sqrt{2}$. The parity transformation

(8.2.3) $$\xi_A \longrightarrow \tilde{\xi}_A \quad , \quad \eta_{A'} \longrightarrow \tilde{\eta}_{A'} \quad ,$$

is given according to the prescription

(8.2.4) $$\tilde{\xi}_A(x) = t^{A'}_A \eta_{A'}(\tilde{x}) \quad , \quad \tilde{\eta}_{A'}(x) = t^A_{A'}\xi_A(\tilde{x})$$

and it is not difficult to verify that the new pair $\tilde{\xi}_A$, $\tilde{\eta}_{A'}$ satisfies the Dirac equation, like the old. Thus, for the Dirac equation, a parity transformation carries us from solution to solution, and is a valid symmetry operation. When electromagnetic interactions are incorporated by putting $\nabla_{AA'} \longrightarrow D_{AA'}$ above with $D_{BB'} = \nabla_{BB'} - i\varepsilon A_{BB'}$, the resulting system is likewise invariant, providing the vector potential $A_{BB'}$ is chosen to transform in the obviously appropriate fashion.

One aspect of the Dirac equation which must be appreciated, as was stressed by Feynman and Gell-Mann (1958), is that it is entirely satisfactory to base the description on a single spinor field ξ_A satisfying the massive wave equation $(\Box + m^2)\xi_A = 0$. The secondary field $\eta_{A'}$ is recoverable as a gradient, using the first "half" of the Dirac equation as given above as a definition (the second "half" following from the wave equation). Similarly, with electromagnetic interaction included, we have the wave equation

$$(D_b D^b + m^2)\xi_A = i\varepsilon\phi_A{}^B\xi_B \quad , \tag{8.2.5}$$

where ϕ_{AB} is the electromagnetic field spinor (i.e., $F_{ab} = \phi_{AB}\varepsilon_{A'B'} + \bar{\phi}_{A'B'}\varepsilon_{AB}$). In each of these cases the parity symmetry is of course still valid, since nothing essential has changed, but the statement of the transformations is somewhat more intricate since a <u>gradient</u> is involved, i.e., $\xi_A \longrightarrow \tilde{\xi}_A$ with $\tilde{\xi}_A(x) = \kappa^{-1} t_A^{A'} \tilde{\nabla}_{A'}^B \xi_B(\tilde{x})$ [substituting $\tilde{\nabla}_{A'}^B$ with $\tilde{D}_{A'}^B$ in the electromagnetic case.] The important fact about weak interaction coupling is that only the left-handed part (i.e., the ξ_A part) of the field figures into the field couplings for particles, and the right-handed (i.e., the $\eta_{A'}$ part) for antiparticles. That is to say, as it is often put, there are no <u>gradient couplings</u> among the fields for weak interactions. Since the parity transformations bring in field gradients in a vital way, as indicated above, it follows that the weak interaction field equations are <u>not</u> invariant under parity transformations. I have purposefully avoided actually writing down these field equations explicitly because they tend to vary, in detail, from version to version of the theory (e.g., with respect to exactly which particle states are being included). But the state of affairs described above has been a central feature in all modern studies of the weak interaction phenomenon. Certain interaction effects are predominantly parity symmetric. This is certainly the case, e.g., with electromagnetism. In that case both parts of the massive field are treated on equal footing, and historically this partially accounts for the present wide-spread usage of the Dirac "four-component spinor algebra", which is so effective because it allows such a symmetry to be systematically incorporated into the formalism. With weak interactions, as we have seen, the circumstances are quite different—there is absolutely no reason, in connection with weak interactions, for not adopting the two-component spinor formalism from the very outset. If the two-component formalism is not employed, the result is simply and utterly grotesque!

8.3 Leptons as Two-Twistor Systems.

Inasmuch as the electron neutrino is always in a left-handed helicity eigenstate, and inasmuch as the left-hand part of the electron wave function plays the "dominant" role in weak interactions, it is convenient to regard these two fields as comprising a left-handed doublet—with this doublet transforming under the action of SU(2). The SU(2) which arises in this connection is somewhat confusingly but not inappropriately called "leptonic isospin", and the $\{\nu_e, e^-\}$ system forms an $I = 1/2$ multiplet, where I denotes the total leptonic isospin quantum number. This approach to classifying leptons does indeed seem to make a great deal of sense, and it forms, for example, the basis of the Weinberg-Salam model, which, on the whole, has been reasonably successful as a lepton model (certainly, the best of all proposals thus far put forward as such) and, in any case, very stimulating as a kind of launching pad for the preparation of more ambitious theories[3]. Inasmuch as the natural internal symmetry group for two-twistor systems contains, as its unitary part, the group U(2), and, seeing how hadrons seem to be describable most naturally as three (or possibly six) twistor systems, and leptons really ought, at least in some sense, to be more elementary than hadrons—although one might argue this point in the case of the τ-particle—it follows that the most logical choice for the representation of leptons, at the outset, is in terms of a pair of twistors[4]. This view, as we shall see, is deficient in at least certain respects.

We wish to describe the $\{\nu_e , e^-\}$ system as a doublet of fields ϕ^i_A . This can be achieved if we introduce a pair of twistors Z^α_i (i = 1, 2) and write

$$\phi^i_A = \oint \rho_x \hat{\pi}^i_A f(Z) \Delta\pi \tag{8.3.1}$$

for the contour integral formula relating the twistor function f(Z) to the fields ϕ^i_A . In order for f(Z) to give rise in a unique way to one of the two fields ϕ^1_A , ϕ^2_A , we require it to be in an eigenstate of an appropriate set of holomorphic differential operators corresponding to leptonic observables.

Let us denote our pair of twistors Z^α_i by

$$Z^\alpha_i = (U^\alpha , D^\alpha) \quad , \tag{8.3.2}$$

in parallel with the hadronic case. The three generators of leptonic isospin are

$$\left\{ \begin{aligned} \hat{I}_1 &= -\tfrac{1}{2}\,(U^\alpha \hat{D}_\alpha + D^\alpha \hat{U}_\alpha) \\ \hat{I}_2 &= -\tfrac{i}{2}\,(U^\alpha \hat{D}_\alpha - D^\alpha \hat{U}_\alpha) \\ \hat{I}_3 &= -\tfrac{1}{2}\,(U^\alpha \hat{U}_\alpha - D^\alpha \hat{D}_\alpha) \quad , \end{aligned} \right. \tag{8.3.3}$$

these formulae being identical, in fact, with equations (6.3.8). The total leptonic isospin operator is

$$\hat{I}^2 = (\hat{I}_1)^2 + (\hat{I}_2)^2 + (\hat{I}_3)^2 \tag{8.3.4}$$

which, when acting on mass eigenstates, is identical (according to Proposition 5.5.9) with the total spin operator. Thus, within a two-twistor framework a leptonic isodoublet <u>must</u> have spin 1/2. The two twistor mass operator is given by

$$\hat{M}^2 = 2\, I_{\alpha\beta} U^\alpha D^\beta I^{\gamma\delta} \hat{U}_\gamma \hat{D}_\delta \quad , \tag{8.3.5}$$

where here, as in (8.3.3), we have written

$$(\hat{U}_\alpha , \hat{D}_\alpha) = \hat{Z}^i_\alpha = -\partial/\partial Z^\alpha_i = (-\partial/\partial U^\alpha , -\partial/\partial D^\alpha) \quad . \tag{8.3.6}$$

Observe that for an electron we must have

(8.3.7)
$$\begin{cases} \hat{M}^2 f(U,D) = m_e{}^2 f(U,D) , \\ \hat{I}^2 f(U,D) = \frac{3}{4} f(U,D) . \end{cases}$$

In addition, we require that f(U,D) be homogeneous of the appropriate degree in U^α and D^α so that only $\emptyset^2_A$ will be non-vanishing in (8.3.1). This is achieved if we put

(8.3.8)
$$\begin{cases} \hat{L} = -U^\alpha \hat{U}_\alpha - D^\alpha \hat{D}_\alpha + 4 , \\ \hat{Q} = D^\alpha \hat{D}_\alpha - 2 , \end{cases}$$

for the operators of <u>lepton number</u>, and <u>electric charge</u>, respectively. Note, incidentally, that we have the formula

(8.3.9)
$$\hat{Q} = \hat{I}_3 - \hat{L}/2 ,$$

which is the leptonic analogue of the Gellmann-Nishijima relation.

The assignments for $\hat{L}$ and $\hat{Q}$ given in equations (8.3.8) are compatible with the contour integral formula

(8.3.10)
$$\emptyset^{A'}_i = \oint \rho_x \pi^{A'}_i f(Z)\Delta\pi$$

for the $\{\bar{\nu}_e , e^+\}$ antiparticle doublet. Note that it is consistent with our observations in Section 8.2 that the particle states described in formula (8.3.1) are left-handed, whereas the antiparticle states described in formula (8.3.10) are automatically right-handed.

A worrying feature of the scheme as it stands is the absence of the muonic leptons. Indeed, within a two-twistor framework there are only two linearly conserved quantum numbers that can be incorporated naturally into the picture. A further drawback arises in connection with the question of the "weak bosons". On account of the leptonic relation $\hat{S}^2 = \hat{I}^2$ we would expect—within a two-twistor framework—a leptonic <u>isotriplet</u> of spin 1 states. However, from phenomenological considerations it would be more consistent to have <u>four</u> states of spin 1 at our disposal, corresponding to the photon, and the three hypothetical weak intermediate bosons W^+, W^-, and Z^o. This difficulty is of the same origin as the ω-ϕ problem described in Section 6.7. One way of resolving the difficulty is to introduce more twistors into the model, an approach which does not seem unreasonable since it allows as well for the additional degrees of freedom needed in order to account for muon number and tau number.

8.4 Models for Sequential Leptons.

Several suggestions have been put forward within the context of twistor theory for lepton models of a more complete character.

One very intriguing approach, which will not be discussed here in any detail, is due to G.A.J. Sparling[5]. He has noted that for massive particles the eigenvalue S(S+1) of the spin-squared operator $-\hat{S}_a\hat{S}^a/m^2$ is invariant under the transformation $S \longrightarrow -S-1$. Therefore he proposes the existence of a new observable $\hat{S}$ which can have two distinct possible eigenvalues for a given value of the spin-squared S(S+1). By exploiting this extra degree of freedom within a twistor context (although no explicit twistorial form for the hypothetical operator $\hat{S}$ has yet emerged) it is possible to build up a lepton model which in at least certain respects seems quite natural. The model predicts heavy leptons, but the resulting spectrum of states differs significantly from the sequential pattern illustrated in Table 8.I. This may turn out to be a desirable feature in the model. But then again, it may not.

Now we turn to the possibilities for building up multiplets of sequential leptons, using a description based on holomorphic functions of several twistors[6].

Let us assume that all the basic leptons come in leptonic isospin doublets, such as the $\{\nu_e , e^-\}$ doublet. Such a doublet can, as we have seen, be represented by a wave function of the form ϕ^{Ai}, with i = 1, 2. Now suppose that we have n such doublets. The first three of these doublets are evidently given by

(8.4.1) $$\{\nu_e , e^-\} \quad , \quad \{\nu_\mu , \mu^-\} \quad , \quad \{\nu_\tau , \tau^-\} \quad ,$$

but one could very well imagine that, at suitably high energies, others could be produced. Let us label these doublets ϕ^{Aip}, with p = 1, 2, ...n. The wave function $\emptyset^{Aip}$ transforms under the action of U(2)xU(n) in a natural way, and describes a multiplet of sequential leptons.

To describe this set-up in twistor terms we introduce a multiplet of 2n "sequential twistors", labeled as follows:

(8.4.2) $$Z^{\alpha}_{ip} \qquad (i = 1, 2 \quad ; \quad p = 1, 2, \ldots n) .$$

Using the notation

(8.4.3) $$Z^{\alpha}_{ip} = (\omega^{A}_{ip} , \pi_{A'ip}) \quad ,$$

(8.4.4) $$\pi^{ip}_{A} = -\partial/\partial\omega^{A}_{ip} \quad ,$$

the multiplet $\emptyset^{Aip}$ can be produced by the following contour integral formula:

(8.4.5) $$\emptyset^{Aip}(x) = \oint \rho_x \hat{\pi}^{Aip} f(Z^{\alpha}_{ip}) \Delta\pi \quad ,$$

and the corresponding antiparticle multiplet is produced in accordance with the formula

(8.4.6) $$\emptyset^{A'}_{ip}(x) = \oint \rho_x \pi^{A'}_{ip} f(Z^{\alpha}_{ip}) \Delta\pi \ .$$

It is not difficult to construct appropriate twistor operator expressions for the observables electric charge, mass, spin, leptonic isospin, and the n sequential lepton numbers (i.e., electron number, muon number, tau number, etc.), and the details of this task can be left to the reader.

As for intermediate vector bosons, it follows on account of the identity

(8.4.7) $$\hat{P}^{AA'} = \hat{\pi}^{Aip} \pi^{A'}_{ip}$$

that there is a single neutral "universally coupled" [i.e., invariant under the action of sequential SU(n)] boson state; and an additional set of 2n-2 "non-universal" neutral bosons, half of which are invariant under leptonic isospin transformations. It is reasonable to suppose that, in the absence of various possible symmetry breaking effects, the single neutral universal boson corresponds to the photon. In the case n=2, there is but a single neutral state invariant under leptonic isospin, and which therefore may couple in an identical fashion to both charged leptons and their neutrino counterparts. This state can be tentatively identified with the Z^o-boson responsible for observed neutral current effects. The role of the remaining neutral bosons (both at the n=2 level, and at higher levels) is not so obvious, nor is it immediately evident what sort of mixing effects involving these states might be expected. In any case, it is worth noting that even if interactions are introduced following the standard patterns of renormalizable field theories (with the vector bosons treated as Yang-Mills fields, and with mass introduced via the Higgs mechanism) the couplings in the resulting model would necessarily differ in certain significant ways from the Weinberg-Salam model, especially insofar as neutral currents are concerned.

Chapter 8, Notes

1. For a very interesting analysis of the discrete symmetries associated with the Poincaré group, see Penrose, 1967, section VII. The corresponding twistor formulae are given there also. The reader should note carefully that the notational conventions vis-a-vis primed and unprimed spinor indices used in Penrose (1967) are opposite to those adopted here.

2. Basic references on the universal Fermi coupling include Feynman and Gell-Mann (1958), and Marshak and Sudarshan (1958).

3. See Weinberg (1967) and (1971).

4. The two-twistor model described here was formulated in collaboration with R. Penrose and G.A.J. Sparling in 1975.

5. See Sparling (1976). The S $\longrightarrow$ -S-1 idea has been pursued further by Penrose, Sparling, and Tsou (1978) in connection with properties of hadrons and Regge trajectories.

6. Discussions with A. Popovich have been very useful in connection with the material of Section 8.4. Cf. Popovich (1978).

CHAPTER 9

SHEAVES AND COHOMOLOGY

9.1 Cochains, Cocycles, and Coboundaries.

The attitude that we have maintained thus far is to regard particle states as functions of one or more twistors, analytic over suitable domains. Since much of the analysis has proceeded more or less at the level of group theory, it has not been necessary to be too specific about the precise nature of these domains—nor has it been necessary to specify the exact nature of the contours involved in the description of n-twistor states. It should be clear, however, that a better understanding of these matters would be quite desirable. Sheaf theory is the key that is required toward this end.

Although it takes some time and practice to get used to the ideas and language of sheaf theory, physicists who are not already acquainted with the subject will certainly find the effort worthwhile. Our approach here will be rather informal. A sheaf over a space M, very crudely speaking, carries the information of the set of all the functions of a particular "type" defined over all the various open sets of M. By a section of a sheaf S over an open state U in M we mean a function (of the "type" with which S is associated) defined on the domain U. The types of functions in which one might be interested are, for example, continuous functions, differentiable functions, holomorphic functions, etc. These might be simply scalar functions—or they might be cross sections of various bundles: e.g., tensor fields. In each case, of course, it is required that M have the structure appropriate for the existence of functions of the relevant "type"—i.e., for holomorphic functions M should be a complex manifold. We refer to S as the "sheaf of germs of functions of type such-and-such". By O we denote, for example, the sheaf of germs of holomorphic functions. The set of all sections of a sheaf S over an open set U forms an abelian group (under addition, in the case of holomorphic functions) and this group is denoted with the following symbol:

$$\Gamma(U,S) \quad . \tag{9.1.1}$$

Now suppose we have a collection of open sets U_i which forms a covering for the space M. If we take a section of S over each of the open sets U_i , then the resulting collection of functions is called a 0-cochain over M, subordinate to the covering U_i . Thus a 0-cochain is an element of the group

(9.1.2) $$C^0(M_{U_i},S) := \Gamma(U_i,S) \quad .$$

In what immediately follows we shall delete the subscript U_i from M, and it will be understood that we are working subordinate to a particular covering. An element of $C^0(M,S)$ can be denoted f_i , where for each value of i , the function f_i is defined on the set U_i .

Now let us write:

(9.1.3) $$\begin{cases} U_{ij} = U_i \cap U_j & , \\ U_{ijk} = U_i \cap U_j \cap U_k & , \\ U_{ijk\ell} = U_i \cap U_j \cap U_k \cap U_\ell & , \text{ etc.,} \end{cases}$$

for the various multiple intersection regions of the sets U_i . The higher cochain groups are then defined as follows:

(9.1.4) $$\begin{cases} C^1(M,S) = \Gamma(U_{ij},S) \\ C^2(M,S) = \Gamma(U_{ijk},S) \\ C^3(M,S) = \Gamma(U_{ijk\ell},S) & , \text{ etc.} \end{cases}$$

Thus, an element of $C^1(M,S)$ is a collection of functions f_{ij} defined over the various double intersection regions U_{ij} . By convention, we always take f_{ij} to be skew-symmetric in its "cohomology indices"; that is to say, we put $f_{ij} = -f_{ji}$. Similarly, an element of $C^q(M,S)$ is given by a collection of functions $f_{ij\ldots k}$ (with q+1 indices) satisfying

(9.1.5) $$f_{ij\ldots k} = f_{[ij\ldots k]} \quad .$$

Let us denote by ρ_i the operator which restricts the domain of whatever it acts on to the intersection of that domain with the set U_i. Thus if f_2 is a function defined on U_2 then $\rho_1 f_2$ is the function defined on U_{12} obtained simply by restricting f_2 down to the subset U_{12}. Now let us consider the map δ defined by

$$(9.1.6)\qquad \begin{cases} C^q(M,S) \xrightarrow{\delta^q} C^{q+1}(M,S) , \\ f_{j\ldots k} \xrightarrow{\delta^q} \rho_{[i} f_{j\ldots k]} . \end{cases}$$

The kernel of δ^q is a subgroup of $C^q(M,S)$ consisting of elements $f_{j\ldots k}$ satisfying

$$(9.1.7)\qquad \rho_{[i} f_{j\ldots k]} = 0 .$$

This relation is known as the <u>cocycle condition</u> and q-cochains satisfying (9.1.7) are called <u>q-cocycles</u>. And thus we define

$$(9.1.8)\qquad Z^q(M,S) = \{f_{j\ldots k} \in C^q(M,S) \mid \rho_{[i} f_{j\ldots k]} = 0\}$$

to be the group of q-cocycles over M (subordinate to the covering U_i) with coefficients in the sheaf S.

The image of δ^q is a subgroup of $C^{q+1}(M,S)$ consisting of all elements $g_{ij\ldots k}$ which happen to be of the special form:

$$(9.1.9)\qquad g_{ij\ldots k} = \rho_{[i} f_{j\ldots k]} ,$$

for some $f_{j\ldots k}$ in $C^q(M,S)$. Elements of this form are called <u>(q+1)-coboundaries</u>. Thus (dropping one notch in q) we define

$$(9.1.10)\qquad B^q(M,S) = \{f_{j\ldots k} \in C^q(M,S) \mid f_{j\ldots k} = \rho_{[j} g_{\ldots k]} , \text{ with } g_{\ldots k} \in C^{q-1}(M,S)\}$$

to be the group of <u>q-coboundaries</u> over M.

Now it should be evident that $B^q(M,S)$ is a subgroup of $Z^q(M,S)$. This is because any $f_{j\ldots k}$ of the form $\rho_{[j} g_{\ldots k]}$ <u>automatically</u> satisfies the cocycle condition (9.1.7). Therefore, we can construct the quotient group

(9.1.11) $$H^q(M,S) = Z^q(M,S)/B^q(M,S) \quad , \quad (q > 0)$$

where two distinct elements of $Z^q(M,S)$ are regarded as equivalent in $H^q(M,S)$ if they differ by a coboundary: that is to say, a pair of cocycles $f_{j...k}$ and $g_{j...k}$ are "representatives" for the same element of $H^q(M,S)$ if and only if there exists an element $\rho_{[j}h_{...k]}$ of $B^q(M,S)$ such that

(9.1.12) $$f_{j...k} = g_{j...k} + \rho_{[j}h_{...k]} \quad .$$

We call $H^q(M,S)$ the qth cohomology group over M (subordinate to the covering U_i) with coefficients in the sheaf S.

The method just outlined gives one of the several available techniques for describing the cohomology groups of a topological space M with coefficients in a sheaf S. The result depends to some extent on the choice of covering U_i, but by resorting to a limiting procedure[(1)] to finer and finer coverings (or, under more auspicious circumstances, simply by choosing U_i sufficiently general) this dependence can be rendered irrelevant.

9.2 Liouville's Theorem, the Laurent Expansion, and the Cohomology of P^1.

The group $H^0(M,S)$ is defined to consist of 0-cochains f_j satisfying the cocycle condition $\rho_{[i}f_{j]} = 0$. This means that whenever two open sets overlap on M the f's defined on these open sets agree with each other on the overlap region. This must hold over the whole space, and so it follows that f_j must be the restriction $\rho_j f$ of global function f to the various open sets U_j . Consequently we have the isomorphism

(9.2.1) $$H^0(M,S) = \Gamma(M,S) \quad ,$$

asserting that $H^0(M,S)$ consists precisely of global sections of the sheaf S over the space M.

Now let us consider the case $M = P^1$, choosing $S = O$, the sheaf of germs of holomorphic functions. Liouville's theorem in complex analysis states that the only analytic functions defined over the entire complex plane, and bounded at infinity, are constants. The condition of boundedness at infinity is equivalent to regularity over the entirety of P^1 (where P^1 is obtained by compactifying the complex plane by adding a point at infinity). Thus Liouville's theorem amounts to the isomorphism

(9.2.2) $$H^0(P^1,O) = C \quad ,$$

where C denotes, as usual, the complex numbers.

What about $H^1(P^1,O)$? Recall that for any analytic function defined in an annular region (i.e., a region excluding, say, the points corresponding to zero and infinity) there exists a Laurent expansion

(9.2.3) $$f(Z) = \frac{1}{2}\left(\sum_{o}^{\infty} a_n Z^n - \sum_{1}^{\infty} b_n Z^{-n}\right) \quad ,$$

splitting f(Z) into an ascending power series and a descending power series. We can think of the annular region as the intersection of two open sets—one including infinity, and the other including zero. Thus using the Laurent expansion any function f_{12} can be split in the form $f_{12} = \rho_{[1}f_{2]}$, where f_1 is analytic throughout the region containing infinity, and f_2 is analytic throughout the region containing zero. Consequently we have[(2)]

(9.2.4) $$H^1(P^1,O) = 0 \quad ,$$

since any cocycle can be expressed as a coboundary, relative to the choice of covering we have made.

More generally one can consider the cohomology on P^1 with coefficients in the sheaf O(n). This sheaf is defined as the sheaf of germs of holomorphic functions $f(\pi_{A'})$, homogeneous of degree n in $\pi_{A'}$, where $\pi_{A'}$ represents the "homogeneous coordinates" on P^1 (i.e., we have $Z = \pi_{0'}/\pi_{1'}$ in equation 9.2.3 above). Note that $f(\pi_{A'})$ is not actually a function on P^1 (except in the case n = 0) but is rather a function[(3)] on the space $C^2-\{0\}$, from which P^1 can be obtained by projection. We can regard O(n) as the sheaf of germs of holomorphic cross-sections of certain bundles (labeled by n) defined over P^1 .

There exists a rather curious connection between the cohomology of P^1 with coefficients in O(n) and spinors. This relationship is very basic, and lies at the heart of both relativity and quantum mechanics. Let us denote by $S^{A'\ldots B'}$ the space of symmetric spinors of valence $[{n \atop 0}]$, and similarly let us write $S_{A'\ldots B'}$ for the dual space. Then it turns out that we have the following isomorphisms:

$$H^{o}(P^{1},O(n)) = S^{A'\ldots B'} \qquad (n \geq 0) \tag{9.2.5}$$

$$H^{o}(P^{1},O(n)) = 0 \qquad (n < 0) \tag{9.2.6}$$

$$H^{1}(P^{1},O(-n-2)) = 0 \qquad (n < 0) \tag{9.2.7}$$

$$H^{1}(P^{1},O(-n-2)) = S_{A'\ldots B'} \qquad (n \geq 0) \tag{9.2.8}$$

Equation (9.2.5) asserts, for example, that the only global sections of $O(n)$ over P^{1} are homogeneous polynomials in $\pi_{A'}$ of degree n; thus a typical representative for an element of $H^{o}(P^{1},O(n))$ would be given by $f^{A'\ldots B'}\pi_{A'}\ldots\pi_{B'}$, where $f^{A'\ldots B'}$ is a constant spinor, with n indices and completely symmetric.

The isomorphism, as dual vector spaces, between $H^{o}(P^{1},O(n))$ and the group $H^{1}(P^{1},O(-n-2))$ is a special case of a more general result of this nature known as Serre duality (Serre, 1955). Let us consider, for example, the case $n = 1$. We want to see how an element of $H^{1}(P^{1},O(-3))$ corresponds to a constant spinor $g_{A'}$ of valence $[{0 \atop 1}]$. Now a representative cocycle for an element of $H^{1}(P^{1},O(-3))$ will be a collection of functions f_{ij} homogeneous of degree -3 in $\pi_{A'}$. If one multiplies f_{ij} by $\pi_{A'}\pi_{B'}$ then the result is homogeneous of degree -1 and, according to (9.2.7), is cohomologically trivial: this means that $\pi_{A'}\pi_{B'}f_{ij}$ is a coboundary, and one can write

$$\pi_{A'}\pi_{B'}\overset{-3}{f}_{ij} = \rho_{[i}\overset{-1}{g}_{j]A'B'} \tag{9.2.9}$$

for some 0-cochain $g_{jA'B'}$ homogeneous of degree -1. Now contracting each side of (9.2.9) with $\pi^{B'}$ one obtains the cocycle condition

$$\rho_{[i}\overset{0}{g}_{j]A'} = 0 \quad , \tag{9.2.10}$$

where we have defined $\overset{0}{g}_{jA'}$ by the formula

$$\overset{0}{g}_{jA'} = \overset{-1}{g}_{jA'B'}\pi^{B'} \quad . \tag{9.2.11}$$

Since $g_{jA'}$ satisfies the cocycle condition it must be the restriction $\rho_j g_{A'}$ of a global function $g_{A'}$, and since $g_{A'}$ is homogeneous of degree zero, it must be constant. And thus we see that we do indeed have an isomorphism as indicated in equation (9.2.8).

9.3 The Cohomology of P^n.

On P^n the set-up is very similar. Let us write $\pi_{a'}$ (a' = 0, ... n) for the homogeneous coordinates on P^n, and write $O(m)$ for the sheaf of germs of holomorphic functions "twisted by m" (i.e.,—homogeneous of degree m on $C^{n+1}-\{0\}$). The only non-vanishing cohomology groups are as follows:

$$H^o(P^n,O(r)) = \text{space of polynomials in } \pi_{a'} \text{ homogeneous of degree r} \tag{9.3.1}$$

$$H^n(P^n,O(-n-1-r)) = \text{dual to } H^o(P^n,O(r)) \quad . \tag{9.3.2}$$

A polynomial homogeneous of degree r will be of the form $f^{a'...b'}\pi_{a'}...\pi_{b'}$, where $f^{a'...b'}$ is a symmetric tensor of valence $[{r \atop o}]$; and thus $H^o(P^n,O(r))$ is the space $S^{a'...b'}$ of such tensors. Similarly, $H^n(P^n,O(-n-1-r))$ is the dual space $S_{a'...b'}$.

Let us consider, as an example, the group $H^2(P^2,O(-3))$. According to our general formulae above we should have an isomorphism $H^2(P^2,O(-3)) = C$. Consequently we should be able to extract the information of a complex number from a representative 2-cocycle f_{ijk}, homogeneous of degree -3. Now since H^2 vanishes for twist > -3 , we must have

$$\pi_{a'}\overset{-3}{f}_{ijk} = \rho_{[i}\overset{-2}{f}_{jk]a'} \tag{9.3.3}$$

for some 1-cochain $\overset{-2}{f}_{jka'}$. Skewing each side of (9.3.3) with $\pi_{b'}$ we have

$$\rho_{[i}\overset{-1}{f}_{jk]a'b'} = 0 \quad , \tag{9.3.4}$$

where $\overset{-1}{f}_{jka'b'}$ is defined by

$$\overset{-1}{f}_{jka'b'} = \overset{-2}{f}_{jk[a'}\pi_{b']} \quad . \tag{9.3.5}$$

Since $f_{jka'b'}$ has twist -1 and satisfies the cocycle condition, it must be cohomologically trivial:

(9.3.6) $$\overset{-1}{f}_{jka'b'} = \rho_{[j}\overset{-1}{f}_{k]a'b'} \quad .$$

Skewing each side with $\pi_{a'}$ and using (9.3.5) we then have

(9.3.7) $$\rho_{[j}\overset{0}{f}_{k]a'b'c'} = 0 \quad ,$$

with $\overset{0}{f}_{ka'b'c'}$ defined by

(9.3.8) $$\overset{0}{f}_{ka'b'c'} = \overset{-1}{f}_{k[a'b'}\pi_{c']} \quad .$$

Since $\overset{0}{f}_{ka'b'c'}$ is global and homogeneous of degree 0 it must be constant. Moreover, since it is skew-symmetric (and a' = 0, 1, 2), it is equivalent to a single complex number f given by the formula $f_{a'b'c'} = f\varepsilon_{a'b'c'}$. It should be evident how this analysis extends to the remaining isomorphisms given in equation (9.3.2). For further discussion see, for example Griffiths and Adams, 1974, pp. 45-55; a proof of the isomorphisms (9.3.1) and (9.3.2) can be found there. These results, established from a somewhat more general point of view, can also be found in Hartshorne, 1977, pp. 225-230.

9.4 The Long Exact Cohomology Sequence.

A sequence of groups and group homomorphisms

(9.4.1) $$\cdots \longrightarrow G_{n-1} \xrightarrow{\alpha_{n-1}} G_n \xrightarrow{\alpha_n} G_{n+1} \longrightarrow \cdots$$

is called exact at G_n if the image of α_{n-1} is precisely the kernel of α_n . If a sequence is exact at each term of the sequence, then we have an exact sequence of groups. An exact sequence of the special form

(9.4.2) $$0 \longrightarrow F \xrightarrow{\alpha} G \xrightarrow{\beta} H \longrightarrow 0$$

is called a short exact sequence. Exact sequences that are longer than short exact sequences are, for reasons which should not be entirely incomprehensible, called long exact sequences. It is not difficult to verify that in order for a sequence (9.4.2) to be exact one must have:

(9.4.3)
$$\begin{cases} \text{(a)} & \alpha \text{ is injective} \\ \text{(b)} & \mathrm{Im}(\alpha) = \mathrm{Ker}(\beta) \\ \text{(c)} & \beta \text{ is surjective} \end{cases}$$

It turns out that it is almost always illuminating to try to formulate problems in terms of the properties of appropriate exact sequences. This goes for physics problems as well as mathematics problems!

In an analogous way one can consider exact sequences of sheaves. A straightforward and very powerful result is available interrelating the cohomology groups associated with a short exact sequence of sheaves:

9.4.4 Proposition. If $0 \longrightarrow R \xrightarrow{\alpha} S \xrightarrow{\beta} T \longrightarrow 0$ is a short exact sequence of sheaves over a space M, then there exists a map δ such that the sequence

(9.4.5)
$$\cdots \longrightarrow H^n(M,R) \xrightarrow{\alpha} H^n(M,S) \xrightarrow{\beta} H^n(M,T) \xrightarrow{\delta} H^{n+1}(M,R) \longrightarrow \cdots$$

is exact.

Proof. We shall simply show the existence of the map δ, and leave the rest of the proof to the reader. Now suppose $f_{j\ldots k}$ is a representative cocycle for an element of $H^n(M,T)$. Since the mapping $S \xrightarrow{\beta} T$ is surjective we can pull back each function in the collection $f_{j\ldots k}$ to form an element $g_{j\ldots k}$ of $C^n(M,S)$. The cocycle condition on $f_{j\ldots k}$ is not necessarily maintained in this procedure, and thus all we can say of $g_{j\ldots k}$ is that it is an element of $C^n(M,S)$. If we now consider $\rho_{[i}g_{j\ldots k]}$ it is interesting to observe that the action of β gives:

(9.4.6)
$$\beta\rho_{[i}g_{j\ldots k]} = \rho_{[i}\beta g_{j\ldots k]} = \rho_{[i}f_{j\ldots k]} = 0 \quad .$$

Since $\rho_{[i}g_{j\ldots k]}$ belongs to the kernel of β it can, on account of the condition $\mathrm{Im}\alpha = \mathrm{Ker}\beta$, be pulled back to some cochain $h_{ij\ldots k}$ in $C^{n+1}(M,R)$, satisfying $\alpha h_{ij\ldots k} = \rho_{[i}g_{j\ldots k]}$. Applying ρ_ℓ and skewing we get

(9.4.7)
$$\alpha\rho_{[\ell}h_{ij\ldots k]} = \rho_{[\ell}\rho_i g_{j\ldots k]} = 0 \quad .$$

However, since α is injective it can only send zero to zero, and thus we must have

(9.4.8) $$\rho_{[\ell}h_{ij\ldots k]} = 0 \quad ,$$

showing that $h_{ij\ldots k}$ is a cocycle. Thus, starting with a representative cocycle in $H^n(M,T)$ we arrive at a representative cocycle in $H^{n+1}(M,R)$. One can easily check that the coboundary freedom works out just right, and thus we have a mapping δ from $H^n(M,T)$ to $H^{n+1}(M,R)$. It is not difficult to verify that δ has properties sufficient to ensure that sequence (9.4.5) is indeed exact[4]. □

The long exact cohomology sequence is of considerable utility when it comes to specific computations, as we shall now proceed to illustrate with several examples.

9.5 The Koszul Complex.

On P^1 let us denote by $O_{A'}(n)$ the sheaf of germs of spinor-valued holomorphic functions, twisted by n. Then we have the following result:

9.5.1 Proposition. The sheaf sequence

(9.5.2) $$0 \longrightarrow O(-2) \xrightarrow{\pi_{A'}} O_{A'}(-1) \xrightarrow{\pi^{A'}} O(0) \longrightarrow 0$$

is exact.

Proof. The map $O(-2) \longrightarrow O_{A'}(-1)$ is injective since $\pi_{A'}$ (homogeneous coordinates for P^1) is non-vanishing. Exactness at $O_{A'}(-1)$ follows from the trivial spinor identity $\pi_{A'}\pi^{A'} = 0$. To prove that $O_{A'}(-1) \longrightarrow O(0)$ is surjective note that if $f(\pi_{A'})$ is a holomorphic function homogeneous of degree zero defined on some small open set U then we can pick a point $\alpha_{A'}$ lying outside of U and write $f_{A'}(\pi_{A'}) = \alpha_{A'}f/\alpha_{B'}\pi^{B'}$, which is non-singular throughout U and homogeneous of degree -1 in $\pi_{A'}$, and satisfies $\pi^{A'}f_{A'} = f$, as required. □

This is a very elementary exact sequence—but nevertheless a sequence of great interest. In fact, when it is jazzed up just a bit it provides (as we shall see) a very direct route for establishing the connection between twistor cohomology and zero rest mass fields. A sequence of slightly greater generality is given by:

$$(9.5.3)\qquad 0 \longrightarrow O(-n-2) \xrightarrow{\overbrace{\pi_{A'}\pi_{B'}\ldots\pi_{C'}}^{n+1}} \underbrace{O_{A'B'\ldots C'}}_{n+1}(-1) \xrightarrow{\pi^{A'}} \underbrace{O_{B'\ldots C'}}_{n}(0) \longrightarrow 0$$

Now if we construct the associated long exact cohomology sequence, the result is

$$(9.5.4)\qquad \ldots \longrightarrow H^0(P^1,O_{A'B'\ldots C'}(-1)) \longrightarrow H^0(P^1,O_{B'\ldots C'}(0)) \longrightarrow$$

$$H^1(P^1,O(-n-2)) \longrightarrow H^1(P^1,O_{A'B'\ldots C'}(-1)) \longrightarrow 0 \ .$$

Thus, providing we know that H^1 and H^0 both vanish for $O(-1)$, we obtain the isomorphism (9.2.8) very directly. To see that $H^0(P^1,O(-1))$ vanishes, note that if $f(\pi_{A'})$ were a global function of $\pi_{A'}$ homogeneous of degree -1 , then $f\alpha_{A'}\pi^{A'}$ would be global and homogeneous of degree 0. Whence by Liouville's theorem $f\alpha_{A'}\pi^{A'} = k$, where k is a constant. But then we would have $f = k/\alpha_{A'}\pi^{A'}$ which blows up at $\alpha_{A'} \sim \pi_{A'}$; thus we have a contradiction unless f simply vanishes. To see that $H^1(P^1,O(-1))$ vanishes, we proceed as follows. Suppose f_{ij} is a representative cocycle for an element of $H^1(P^1,O(-1))$. Multiplying by $\pi_{A'}$ we get

$$(9.5.5)\qquad \pi_{A'}\overset{-1}{f}_{ij} = \rho_{[i}\overset{0}{f}_{j]A'}$$

for some $\overset{0}{f}_{jA'}$, since we know $H^1(P^1,O(0))$ is trivial. Transvecting with $\pi^{A'}$ we get $\rho_{[i}f_{j]A'}\pi^{A'} = 0$, showing that $f_{jA'}\pi^{A'}$ is global:

$$(9.5.6)\qquad f_{jA'}\pi^{A'} = \rho_j\alpha_{A'}\pi^{A'} \quad , \quad (\alpha_{A'} \text{ constant}) \quad .$$

Equation (9.5.6) implies that $f_{jA'}$ must be of the form:

$$(9.5.7)\qquad f_{jA'} = \rho_j\alpha_{A'} + \overset{-1}{f}_j\pi_{A'} \quad .$$

Substituting (9.5.7) in (9.5.5), we get the desired result, namely: $\pi_{A'}\overset{-1}{f}_{ij} = \rho_{[i}\overset{-1}{f}_{j]}\pi_{A'}$, showing that f_{ij} is indeed cohomologically trivial.

A result quite analogous to Proposition 9.5.1 holds for P^n. Let us denote homogeneous coordinates on P^n, as before, by $\pi_{a'}$. And we shall write $O_{a'\ldots b'}(r)$ for the sheaf of germs of skew-symmetric tensor valued holomorphic functions, twisted by r. Then we have:

9.5.8 Proposition. The sequence

$$(9.5.9) \qquad 0 \longrightarrow O(-r-1) \longrightarrow O_{a'}(-r) \longrightarrow O_{a'b'}(-r+1) \longrightarrow \cdots$$

$$\cdots \longrightarrow O_{a'b'\ldots c'}(0) \longrightarrow 0$$

is exact, where in each case the sheaf maps are given by multiplication by $\pi_{a'}$ and skew-symmetrizing in an appropriate fashion.

The proof is quite analogous to that of Proposition 9.5.1. It is not difficult, also, to construct an analog of sequence (9.5.3). Then by applying the exact cohomology sequence one can obtain directly the isomorphisms (9.3.1) and (9.3.2). Sequence (9.5.9) is a special example of what is known as the Koszul complex. As will be discussed in Section 10.6, it plays a special role in the analysis of the cohomology of functions of several twistors.

9.6 Line Bundles and Chern Classes.

It is easy to see that the set of all nowhere-vanishing holomorphic functions on a region U forms a group under multiplication. The corresponding sheaf is denoted O^*. Since locally any nowhere-vanishing holomorphic function g can be expressed in the form $\exp(f) = g$, where f is a holomorphic function, the following sequence is exact, where Z denotes the integers:

$$(9.6.1) \qquad 0 \longrightarrow Z \longrightarrow O \longrightarrow O^* \longrightarrow 0 \quad ,$$

where the map $Z \longrightarrow O$ is simply multiplication by $2\pi i$.

The cohomology sequence associated with (9.6.1) contains the segment:

$$(9.6.2) \qquad \cdots \longrightarrow H^1(M,O) \longrightarrow H^1(M,O^*) \longrightarrow H^2(M,Z) \longrightarrow \cdots \quad .$$

The group $H^1(M,O^*)$ is called the group of holomorphic line bundles over M, and each element of $H^1(M,O^*)$ is called a <u>line bundle</u>. A line bundle is specified by giving a representative cocycle ξ_{ij} , which must satisfy the cocycle condition

$$\rho_{ijk}\xi_{ij}\xi_{jk}\xi_{ki} = 1 \quad , \tag{9.6.3}$$

where ρ_{ijk} denotes restriction to the triple intersection region U^{ijk} . Note that (9.6.3) is satisfied trivially if we put

$$\xi_{ij} = \rho_{ij}\xi_i/\xi_j \quad , \tag{9.6.4}$$

where ξ_i is a collection of nowhere-vanishing holomorphic functions defined over U_i. Thus the coboundary freedom available in the specification of a line bundle is given by

$$\xi_{ij} \longrightarrow \rho_{ij}\xi_i\xi_{ij}\xi_j^{-1} \quad . \tag{9.6.5}$$

The element of $H^2(M,Z)$ to which a line-bundle ξ_{ij} is mapped in (9.6.2) is called the <u>Chern class</u> of the line bundle. From the exactness of (9.6.2) it should be evident that line bundles with vanishing Chern class are precisely those which can be obtained by "exponentiating" elements of $H^1(M,O)$; i.e., a line bundle ξ_{ij} has vanishing Chern class if and only if it can be expressed in the form

$$\xi_{ij} = \exp(f_{ij}) \quad , \tag{9.6.6}$$

with f_{ij} satisfying the additive cocycle relation

$$\rho_{[i}f_{jk]} = 0 \quad . \tag{9.6.7}$$

The notion of line bundle is a special case of the notion of a <u>vector bundle</u>[5] over a space M. A holomorphic vector bundle is defined to be an element of the group $H^1(M,O^a_b)$, where O^a_b is the sheaf of holomorphic non-singular matrix valued functions. An element of $H^1(M,O^a_b)$ is specified by a collection ξ^a_{ijb} of such functions over U_{ij} satisfying

$$\rho_{ijk}\xi^a_{ijb}\xi^b_{jkc}\xi^c_{kid} = \delta^a_d \qquad (\text{in } U_{ijk}) \quad . \tag{9.6.8}$$

9.7 Varieties, Syzygies, and Ideal Sheaves.

A projective algebraic variety is defined to be the common zero set of a collection f_r of homogeneous polynomials in the homogeneous coordinates Z^a of complex projective n-space. If there is but a single homogeneous polynomial f, then the variety V defined by f = 0 is called a hypersurface of degree q, where q is the degree of the polynomial f. Hypersurfaces of degree q = 1, 2, 3, 4, ... are called hyperplanes, quadrics, cubics, quartics, etc., respectively.

As a simple example of an algebraic variety one can consider the embedding of $P^1 x P^1$ as a quadric hypersurface in P^3. Suppose we write $\pi_i^{A'}$ (A' = 1, 2; i = 1, 2) for the four homogeneous coordinates of P^3. Then the quadratic equation

$$\pi_i^{A'} \pi_j^{B'} \varepsilon_{A'B'} \varepsilon^{ij} = 0 \tag{9.7.1}$$

has the solution

$$\pi_i^{A'} = \pi^{A'} \lambda_i \quad . \tag{9.7.2}$$

The variables $\pi^{A'}$ and λ_i (which are determined by 9.7.2 only up to scale) serve as homogeneous coordinates for $P^1 x P^1$.

As a somewhat more complicated example, let us consider the embedding of $P^1 x P^2$ as an algebraic variety in P^5. For homogeneous coordinates on P^5 let us write $\pi_{iA'}$ (i = 1, 2, 3; A' = 1, 2). Then $P^1 x P^2$ is given by the locus

$$\pi_{iA'} \pi_{jB'} \varepsilon^{A'B'} = 0 \quad , \tag{9.7.3}$$

for which the solution is

$$\pi_{iA'} = \pi_{A'} \lambda_i \quad , \tag{9.7.4}$$

with $\pi_{A'}$ and λ_i acting as homogeneous coordinates for P^1 and P^2, respectively. It should be noted that the three equations (9.7.3) are not completely independent, since we have the relations

$$f_{ij} \pi_{kA'} \varepsilon^{ijk} = 0 \quad , \tag{9.7.5}$$

which are satisfied automatically, where f_{ij} is defined by

(9.7.6) $$f_{ij} = \pi_{iA'}\pi_{jB'}\varepsilon^{A'B'} \quad .$$

Associated with any projective algebraic variety V is an ideal sheaf I_V , defined to be the sheaf of germs of holomorphic functions which vanish on V. The ideal sheaf can be described by an exact sequence

(9.7.7) $$0 \longrightarrow I_V \longrightarrow O \xrightarrow{\rho_V} O_V \longrightarrow 0 \quad ,$$

where ρ_V is the restriction map down to the variety, and O_V is the sheaf of germs of holomorphic functions defined on the variety. In the case of $P^1xP^1 \subset P^3$ we observe that, locally, any holomorphic function which vanishes when restricted down from its domain in P^3 to the intersection of that domain with the quadratic P^1xP^1 must be of the form $\pi_i^{A'}\pi_j^{B'}\varepsilon^{ij}\varepsilon_{A'B'}f(\pi_i^{A'})$, where $f(\pi_i^{A'})$ is homogeneous of degree -2. Thus, in this case (9.7.7) can be written more explicitly in the form

(9.7.8) $$0 \longrightarrow O(-2) \longrightarrow O_{P^3} \longrightarrow O_{P^1xP^1} \longrightarrow 0 \quad .$$

In the case of $P^1xP^2 \subset P^5$ the syzygy (9.7.5) plays a role, and the analog of sequence (9.7.8) is a long exact sequence. This is because any function on a region of P^5 which vanishes when restricted down to P^1xP^2 must be of the form $g^{ij}f_{ij}$, with f_{ij} as defined in (9.7.6) and g^{ij} an arbitrary holomorphic function (twisted by -2). Therefore it follows that the sequence

(9.7.9) $$O^{ij}(-2) \longrightarrow O_{P^5} \longrightarrow O_{P^1xP^2} \longrightarrow 0$$

is exact. However, we can substitute

(9.7.10) $$g^{ij} \longrightarrow g^{ij} + \varepsilon^{ijk}\pi_{A'k}h^{A'}(\pi) \quad ,$$

where $h^{A'}(\pi)$ is an arbitrary function homogeneous of degree -3, and leave $g^{ij}f_{ij}$ invariant. Thus, on account of (9.7.5) we obtain a long exact sequence [6]:

(9.7.11) $$0 \longrightarrow O^{A'}(-3) \longrightarrow O^{ij}(-2) \longrightarrow O_{P^5} \longrightarrow O_{P^1xP^2} \longrightarrow 0 \quad .$$

Consequently, since the first three of the sheaves in (9.7.11) are defined on P^5, the cohomology of P^1xP^2 can be related to various cohomology groups defined on P^5 (using the long exact cohomology sequence). Of course in this case we can compute the cohomology of P^1xP^2 directly by other means; but for other varieties (which may not have the pleasure of being products of projective spaces) we can construct analogs of (9.7.11) and reduce the problem of computing the cohomology of V to an elementary problem in linear algebra.

For example, suppose one is interested in the cohomology of a non-singular cubic surface $A_{\alpha\beta\gamma}Z^{\alpha}Z^{\beta}Z^{\gamma} = 0$ in P^3. In this case the relevant exact sequence is

(9.7.12) $$0 \longrightarrow O_{P^3}(n-3) \xrightarrow{\alpha} O_{P^3}(n) \longrightarrow O_V(n) \longrightarrow 0 \quad ,$$

where $O_V(n)$ is the sheaf of germs of holomorphic functions on the cubic surface, twisted by n, and the map α is multiplication by the function $A_{\alpha\beta\gamma}Z^{\alpha}Z^{\beta}Z^{\gamma}$. Taking the cohomology of (9.7.12), a short calculation gives

(9.7.13) $$H^0(V,O_V(n)) = \mathrm{Coker}(\alpha^*) \quad , \quad \alpha^*:H^0(P^3,O(n)) \longrightarrow H^0(P^3,O(n-3)) \quad ,$$

and

(9.7.14) $$H^2(V,O_V(n)) = \mathrm{Ker}(\alpha^*) \quad , \quad \alpha^*:H^3(P^3,O(n-3)) \longrightarrow H^3(P^3,O(n)) \quad .$$

After a little thought one will recognize (9.7.13) as the space of symmetric dual twistors $P_{\alpha\beta\gamma\delta\ldots\varepsilon}$ of valence n, modulo the space of symmetric valence n dual twistors of the special form $A_{(\alpha\beta\gamma}P_{\delta\ldots\varepsilon)}$, where $P_{\delta\ldots\varepsilon}$ is of valence n-3. In (9.7.14) one finds the dual space to (9.7.13), this being the space of symmetric twistors $P^{\alpha\beta\gamma\delta\ldots\varepsilon}$ of valence n which are <u>annihilated</u> by $A_{\alpha\beta\gamma}$:

(9.7.15) $$P^{\alpha\beta\gamma\delta\ldots\varepsilon}A_{\alpha\beta\gamma} = 0 \ .$$

As another example, let us consider the so-called "twisted cubic" curve in P^3. In this case it is most convenient to parameterize P^3 by symmetric spinors of valence 3. Thus, the four independent components of ξ_{ABC} act as homogeneous coordinates for P^3. The twisted cubic curve is defined to be the locus

$$\xi_{ABC}\xi^{C}_{EF} = 0 \quad . \tag{9.7.16}$$

Writing $\emptyset_{ABEF} = \xi_{ABC}\xi^{C}_{EF}$ it is straightforward to verify the property

$$\emptyset_{ABEF} = -\,\emptyset_{EFAB} \quad . \tag{9.7.17}$$

Thinking of the symmetric index pairs AB and EF as index clumps, we see that $\emptyset_{ABEF}$ is a skew-symmetric three-by-three matrix, and accordingly has three essential components. Thus (9.7.16) represents the intersection of three quadrics. This gives us the exact sequence

$$O_{P^3}^{ABEF}(n-2) \xrightarrow{\emptyset_{ABEF}} O_{P^3}(n) \longrightarrow O_T(n) \quad , \tag{9.7.18}$$

where $O_T(n)$ is the sheaf of germs of holomorphic functions on the twisted cubic, twisted by n. In order to continue (9.7.18) to form a long exact sequence, we need the elementary spinor identity

$$\emptyset_{ABEF}\xi^{ABE} = 0 \quad . \tag{9.7.19}$$

This gives us the sequence

$$0 \longrightarrow O^A(n-3) \xrightarrow{\alpha} O^{ABEF}(n-2) \longrightarrow O(n) \longrightarrow O_T(n) \longrightarrow 0 \quad , \tag{9.7.20}$$

where the sheaf map α is specified by

$$O^A \longrightarrow O^{(A}\xi^{B)EF} - O^{(E}\xi^{F)AB} \quad , \tag{9.7.21}$$

in order to ensure that the image has the correct symmetries. Given the exact sequence (9.7.20), it will be left to the reader to work through the details of sorting out the associated long exact cohomology sequence, this being intricate but not difficult.

This concludes our brief introduction to sheaves and cohomology. All of the material mentioned here is useful in one way or another in connection with twistor theory, although not all that has been said will be used in the next chapter. For further material the reader is referred to Serre (1956), Gunning and Rossi (1965), Gunning (1966), Chern (1967), Morrow and Kodaira (1971), Godement (1973),

Shafarevich (1977), Hartshorne (1977), and numerous other references. It should be stressed that there are many intimate interconnections between quantum mechanics and the theory of algebraic varieties—it is reasonable to speculate, in fact, that all the discrete degrees of freedom that manifest themselves in quantum mechanics can be understood ultimately in terms of the cohomology of algebraic varieties. For the various continuous degrees of freedom that appear in quantum mechanics, however, it would appear that more general categories of complex manifolds (i.e., non-algebraic manifolds) must be investigated.

Chapter 9, Notes

1. For a description of the limiting procedure involved here see Gunning, 1966, p. 30.. Also, see pp. 44-47 in the same reference for a discussion of "Leray's theorem" which gives a set of conditions sufficient to ensure that a covering U_i is general enough to calculate the cohomology of a space M.

2. Strictly speaking in order to establish this result we need to know that a covering of P^1 by two open sets suffices to compute its cohomology.

3. Cross-sections of the sheaf $O(n)$ are often referred to as "twisted functions"; and $O(n)$ itself is called the "sheaf of germs of holomorphic functions, twisted by n".

4. For further discussion of the long exact cohomology sequence, see, for example, Gunning, 1966, pp. 32-34.

5. Holomorphic line bundles and holomorphic vector bundles—built over suitable regions of projective twistor space—can be used to describe self-dual solutions of Maxwell's equations and the Yang-Mills equations (without sources). See Ward (1977a and 1977b), Atiyah and Ward (1977), Hartshorne (1978), and Ward (1979) for various details of the procedure. Also see Burnett-Stuart (1978) and Moore (1978).

6. Note that for sequence (9.7.8) we have an isomorphism between $O(-2)$ and I_V. In the case of sequence (9.7.11) we have the following isomorphism:

$$I_V = O^{ij}(-2)/\text{Image}[O^{A'}(-3)] \quad .$$

CHAPTER 10

APPLICATIONS OF COMPLEX MANIFOLD TECHNIQUES TO ELEMENTARY PARTICLE PHYSICS

10.1 The Kerr Theorem.

Standing before us we see two alternative pictures of reality. On the one hand there is spacetime, and on the other there is twistor space. Einstein has taught us that gravitation is itself but an aspect of the structure of spacetime. According to the view of twistor theory, gravitation is to be reinterpreted in terms of the complex analytic geometry of twistor space. Elementary particle states are to be interpreted similarly—in fact, according to Penrose we are to think ultimately of actually in some sense incorporating elementary particle states directly "into" the complex analytic structure of twistor space.

At the moment only a few examples of this procedure are known in sufficiently explicit detail to make comment worthwhile—however, there is no reason to suppose that these techniques cannot be generalized to accommodate a reasonable spectrum of particles, and to treat certain features of their interactions as well.

At the mention of interactions what springs to mind immediately is the question of how the various non-linearities of field theory are to be realized in complex analytic terms. The Kerr theorem provides a striking illustration of the fact that certain non-linear partial differential equations arising in connection with properties of fields on Minkowski space can be reinterpreted in a very straightforward way in terms of the complex analytic geometry of twistor space. The Kerr theorem has its origin in certain special classes of Maxwell's equations, called <u>null electromagnetic fields</u>. A null electromagnetic field is a solution of Maxwell's equations for which both of the invariants $F_{ab}F^{ab}$ and $*F_{ab}F^{ab}$ vanish. Equivalently, if the electromagnetic field spinor $\phi_{A'B'}$ is introduced according to the familiar scheme

(10.1.1) $$F_{ab} = \phi_{A'B'}\varepsilon_{AB} + \bar{\phi}_{AB}\varepsilon_{A'B'} \quad ,$$

then F_{ab} is null if and only if $\phi_{A'B'}$ is of the form

(10.1.2) $$\phi_{A'B'} = \phi\pi_{A'}\pi_{B'} \quad ,$$

for some choice of ϕ and $\pi_{A'}$. According to a remarkable theorem of Robinson (1959), if a spinor field $\pi_{A'}$ satisfies the geodesic shearfree condition

(10.1.3) $$\pi^{A'}\pi^{B'}\nabla_{AA'}\pi_{B'} = 0 \ ,$$

then there will always exist a choice of ϕ such that $\phi_{A'B'}$, as defined in (10.1.2), satisfies Maxwell's equations $\nabla^{AA'}\phi_{A'B'} = 0$; and conversely, if a spinor field $\phi_{A'B'}$ satisfies Maxwell's equations and is of the form (10.1.2), then $\pi_{A'}$ satisfies (10.1.3). Equation (10.1.3) asserts that $\pi_{A'}$ is tangent to a family of geodesic shearfree null rays. According to the theorem of Kerr[(1)], such geodesic shearfree congruences can be characterized in terms of complex analytic surfaces in P^3.

A complex analytic surface in P^3 is defined by the vanishing of a holomorphic function $f(Z^\alpha)$, homogeneous of some degree n. If $f(Z^\alpha)$ should happen to be a homogeneous polynomial, then the surface $f(Z^\alpha) = 0$ [which will henceforth be denoted S] is an algebraic surface; but more generally we simply have an analytic surface.

Let us denote by X a complex projective line in P^3 corresponding to a space-time point $x^{AA'}$. Now if Z^α is an intersection point of X and S, then Z^α must be of the form

(10.1.4) $$Z^\alpha = (ix^{AA'}\pi_{A'} \ , \ \pi_{A'}) \quad ,$$

and must satisfy

(10.1.5) $$f(ix^{AA'}\pi_{A'} \ , \ \pi_{A'}) = 0 \quad .$$

If we vary the line X, then $\pi_{A'}$ must be correspondingly adjusted if (10.1.5) is to remain satisfied. In this way we obtain a field of spinors $\pi_{A'}(x)$, determined up to proportionality, satisfying

(10.1.6) $$f[ix^{AA'}\pi_{A'}(x) \ , \ \pi_{A'}(x)] = 0 \quad .$$

In general the field $\pi_{A'}(x)$ will possess several distinct "branches", since it is possible for a given line X to intersect S in more than one place.

10.1.7 Theorem. If a spinor field $\pi_{A'}(x)$ satisfies (10.1.6) for some holomorphic function $f(Z^\alpha)$, then it satisfies equation (10.1.3).

Proof. Since $f(Z^\alpha)$ is homogeneous of degree n we have

$$Z^\alpha \frac{\partial f}{\partial Z^\alpha} = nf \quad , \tag{10.1.8}$$

whence on the surface S we have

$$Z^\alpha \frac{\partial f}{\partial Z^\alpha} = \omega^A \frac{\partial f}{\partial \omega^A} + \pi_{A'} \frac{\partial f}{\partial \pi_{A'}} = 0 \ . \tag{10.1.9}$$

Then, restricting to the intersection of S with X, we obtain

$$ix^{BB'}\pi_{B'} \frac{\partial f}{\partial \omega^B} + \pi_{B'} \frac{\partial f}{\partial \pi_{B'}} = 0 \quad , \tag{10.1.10}$$

which implies the existence of a scalar λ such that

$$ix^{BB'} \frac{\partial f}{\partial \omega^B} + \frac{\partial f}{\partial \pi_{B'}} = \lambda\, \pi^{B'} \quad . \tag{10.1.11}$$

Storing this bit of information, let us return to equation (10.1.6). Since (10.1.6) must remain valid if we vary $x^{AA'}$, it follows that the derivative

$$\nabla_{AA'} f[ix^{BB'}\pi_{B'}(x) \ , \ \pi_{B'}(x)] \tag{10.1.12}$$

must vanish. With a straightforward application of the chain rule, the vanishing of (10.1.12) implies

$$i\pi_{A'} \frac{\partial f}{\partial \omega^A} + (\nabla_{AA'}\pi_{B'})[ix^{BB'} \frac{\partial f}{\partial \omega^B} + \frac{\partial f}{\partial \pi_{B'}}] = 0 \quad . \tag{10.1.13}$$

Transvecting (10.1.13) with $\pi^{A'}$ and using (10.1.11), the desired result (10.1.3) follows immediately. □

As an example of the Kerr theorem at work, let us consider again a cubic surface in P^3, given by the equation

$$A_{\alpha\beta\gamma} Z^\alpha Z^\beta Z^\gamma = 0 \quad . \tag{10.1.14}$$

Now associated with $A_{\alpha\beta\gamma}$ is a solution $\xi^{A'B'C'}(x)$ of the equation

(10.1.15) $$\nabla^{A(A'}\xi^{B'C'D')} = 0 \quad ,$$

where $\xi^{A'B'C'}$ is defined by

(10.1.16) $$\rho_x A_{\alpha\beta\gamma} Z^\alpha Z^\beta Z^\gamma = \xi^{A'B'C'} \pi_{A'}\pi_{B'}\pi_{C'} \quad ,$$

with $\rho_x Z^\alpha = (ix^{AA'}\pi_{A'}\ ,\ \pi_{A'})$, as usual. Thus, if we put

(10.1.17) $$Z^\alpha = [ix^{AA'}\pi_{A'}(x)\ ,\ \pi_{A'}(x)]$$

in equation (10.1.14), then we obtain

(10.1.18) $$\xi^{A'B'C'}\pi_{A'}(x)\pi_{B'}(x)\pi_{C'}(x) = 0$$

as the formula for our shearfree congruence. Equation (10.1.18) has three distinct solutions, for a general cubic surface, and these are given by

(10.1.19) $$\pi_{A'} \approx \alpha_{A'} \quad , \quad \pi_{A'} \approx \beta_{A'} \quad , \quad \pi_{A'} \approx \gamma_{A'} \quad ,$$

where $\pi_{A'}$, $\beta_{A'}$, and $\gamma_{A'}$ are the principal spinors of $\xi^{A'B'C'}$:

(10.1.20) $$\xi^{A'B'C'} = \alpha^{(A'}\beta^{B'}\gamma^{C')} \quad .$$

It is not difficult to verify that as a consequence of (10.1.15) all three of these principal spinors satisfy the geodesic shearfree condition, and thus define the three branches of the congruence of geodesic shearfree rays associated with the cubic surface (10.1.14). Incidentally, according to a classical result of projective geometry due to Schläfli, a cubic surface in P^3 has exactly 27 lines lying on it—in spacetime this result corresponds to the fact that a solution of (10.1.15) must possess exactly 27 zero-points (i.e., points where $\xi_{A'B'C'}$ vanishes). A detailed investigation of the geometry of this configuration would undoubtedly make for a highly amusing exercise(2).

10.2 Zero Rest Mass Fields as Elements of Sheaf Cohomology Groups.

Now we come to the question of analyzing zero rest mass fields from the perspective of twistor sheaf cohomology. We are interested here in the cohomology group $H^1(M,O(n))$, where M is a region of P^3 , and where $O(n)$ is the sheaf of germs of holomorphic functions twisted by n. The sort of region M in which we are interested is one that is swept out by a set of projective lines in P^3 corresponding to a set of points in complex Minkowski space. For positive frequency fields, for example, the region of complex Minkowski space of interest is CM^+ , and the corresponding region in P^3 is PT^+ .

<u>10.2.1 Proposition</u>. Each element of $H^1(M,O(n))$ corresponds to a zero rest mass field of helicity $s = -\frac{1}{2}n-1$ defined over the region of complex Minkowski space to which M is related. Distinct elements correspond to distinct zero rest mass fields.

<u>Proof</u>. First let us consider the case $s > 0$, i.e., $n < -2$. In order to simplify the discussion we shall examine the case $s = 1/2$ explicitly, and the reader should have no difficulty in filling in the details required for higher helicities.

Suppose that f_{ij} is a representative cocycle for an element of $H^1(M,O(-3))$. If we restrict f_{ij} down to the complex projective line P^1_x corresponding to a space-time point $x^{AA'}$, then $\rho_x f_{ij}$ can be regarded as a function of $x^{AA'}$ and $\pi_{A'}$. For fixed $x^{AA'}$ then $\rho_x f_{ij}$ is a representative cocycle of $H^1(P^1_x,O(-3))$. Now we can apply the analysis of Section 9.2. If we multiply $\rho_x f_{ij}$ by $\pi_{A'}\pi_{B'}$, the result must be cohomologically trivial, so we get

(10.2.2) $$\pi_{A'}\pi_{B'}\rho_x \overset{-3}{f}_{ij} = \rho_{[i}\overset{-1}{g}_{j]A'B'}$$

for some 0-cochain $\overset{-1}{g}_{jA'B'}$, which is a function of $x^{AA'}$ and $\pi_{A'}$. Transvecting (10.2.2) with $\pi^{A'}$ we get (cf. equation 9.2.9):

(10.2.3) $$\rho_{[i}\overset{0}{\emptyset}_{j]A'} = 0 \quad ,$$

where $\overset{0}{\emptyset}_{jA'}$ is defined by

(10.2.4) $$\overset{0}{\emptyset}_{jA'} = \overset{-1}{g}_{jA'B'}\pi^{B'} \quad .$$

Since $\overset{0}{\emptyset}_{jA'}$ is global and homogeneous of degree zero in $\pi_{A'}$, it must be <u>constant</u>

in $\pi_{A'}$, and thus a function of $x^{AA'}$ alone, with

(10.2.5) $$\emptyset_{jA'} = \rho_j \emptyset_{A'}(x) \quad .$$

To prove that $\emptyset_{A'}(x)$ satisfies the zero rest mass equations we must note that since f_{ij} is a collection of twistor functions it must satisfy

(10.2.6) $$\pi_{A'} \nabla_A^{A'} \rho_x f_{ij} = 0 \quad .$$

Consequently, if we transvect (10.2.2) with $\nabla^{A'A}$ we get:

(10.2.7) $$\rho_{[i} \nabla^{AA'} \rho_x g_{j]A'B'} = 0 \quad ,$$

which says that $\nabla^{AA'} \rho_x g_{jA'B'}$ is global. However, since $\nabla^{AA'} \rho_x g_{jA'B'}$ is homogeneous of degree -1 in $\pi_{A'}$, the only way it can be global is for it to vanish:

(10.2.8) $$\nabla^{AA'} \rho_x g_{jA'B'} = 0 \quad .$$

Transvecting (10.2.8) with $\pi^{B'}$ and using (10.2.4) and (10.2.5), the desired result $\nabla^{AA'} \emptyset_{A'} = 0$ follows immediately.

Next, we must prove that $\emptyset_{A'}$ is independent of the coboundary freedom available in the specification of f_{ij} . To see this, observe that the transformation

(10.2.9) $$\overset{-3}{f}_{ij} \longrightarrow \overset{-3}{f}_{ij} + \rho_{[i} \overset{-3}{g}_{j]}$$

must be accompanied by the substitution

(10.2.10) $$\overset{-1}{g}_{jA'B'} \longrightarrow g_{jA'B'} + g_j \pi_{A'} \pi_{B'}$$

in equation (10.2.2). However, a glance at (10.2.4) shows that $\emptyset_{jA'}$ is invariant under this substitution. Conversely, we wish to see that if $\emptyset_{A'}$ vanishes, then f_{ij} must be cohomologically trivial. From (10.2.4) it follows that if $\emptyset_{A'}$ vanishes, then $g_{jA'B'}$ must be of the form $g_j \pi_{A'} \pi_{B'}$ for some 0-cochain g_j dependent upon $x^{AA'}$ and $\pi_{A'}$. From (10.2.2) it then follows that

(10.2.11) $$\rho_x \overset{-3}{f}_{ij} = \rho_{[i} \overset{-3}{g}_{j]} \quad ,$$

but we are not done yet, since it remains to be shown that g_j is indeed a 0-cochain on twistor space—thus far, we have merely established that g_j is dependent on $x_{AA'}$ and $\pi_{A'}$. The situation is immediately remedied, however, if we hit (10.2.11) with $\nabla_{A'}\pi_A^{A'}$, thereby obtaining

$$\rho_{[i}\pi_{A'}\nabla_A^{A'}\overset{-3}{g}_{j]} = 0 \quad , \tag{10.2.12}$$

which implies, since $\pi_{A'}\nabla_A^{A'}g_j$ is homogeneous of degree -2, that $\pi_{A'}\nabla_A^{A'}g_j$ vanishes.

That concludes the proof for s = 1/2. Now let us consider the case s = - 1/2. If f_{ij} is homogeneous of degree -1, then $\rho_x f_{ij}$ is cohomologically trivial, i.e., we have:

$$\rho_x\overset{-1}{f}_{ij} = \rho_{[i}\overset{-1}{g}_{j]} \quad , \tag{10.2.13}$$

for some 0-cochain $\overset{-1}{g}_j$. If we operate on (10.2.13) with $\pi_{A'}\nabla_A^{A'}$ then we obtain

$$\rho_{[i}\pi_{A'}\nabla_A^{A'}\overset{-1}{g}_{j]} = 0 \tag{10.2.14}$$

which implies, since $\pi_{A}\nabla^{A'}_{A}\overset{-1}{g}_j$ is homogeneous of degree zero, that

$$\pi_{A'}\nabla_A^{A'}\overset{-1}{g}_j = \rho_j\phi_A(x) \quad , \tag{10.2.15}$$

where $\phi_A(x)$ is a function of $x^{AA'}$ alone. Transvecting equation (10.2.15) with $\pi_{B'}\nabla^{B'A}$ we get

$$\pi_{B'}\nabla^{B'A}\phi_A = 0 \tag{10.2.16}$$

on account of the identity $\nabla^A_{(A'}\nabla_{B')A} = 0$. And since (10.2.16) must hold for all values of $\pi_{A'}$, we get the field equation $\nabla^{A'A}\phi_A = 0$, as desired. It is straightforward to check that ϕ_A is independent of the coboundary freedom available to f_{ij}. Moreover, one can verify that if ϕ_A vanishes, then f_{ij} itself must be cohomologically trivial. □

10.3 Spin-Bundle Sequences.

The results of Section 10.2 can be obtained from a somewhat more refined point of view through the consideration of various exact sequences of sheaves. It should be pointed out that quite a bit of lore has already been developed in this connection, and it would be impossible to give an account here which is in any sense inclusive of all the work that has been done in this vein. To simplify matters, let us assume that we are working away from infinity in complex Minkowski space, and thus that the region of twistor space with which we are concerned is $P^3 - P^1_I$, where P^1_I is the line $I^{\alpha\beta}$. Since P^1_I is given by the equation $\pi_{A'} = 0$, we can take $\pi_{A'}$ now to be non-vanishing. As before, we shall denote by $O(n)$ the sheaf of germs of holomorphic functions on twistor space, twisted by n. Let us denote by $F(n)$ the sheaf of germs of functions of $x^{AA'}$ and $\pi_{A'}$, homogeneous of degree n in $\pi_{A'}$. The sheaf $F(1)$ is called the sheaf of germs of holomorphic cross sections of the <u>spin bundle</u>. The subsheaf of $F(n)$ consisting of germs $f(x,\pi)$ satisfying $\pi_{A'}\nabla^{A'}_A f(x,\pi) = 0$ is naturally isomorphic to $O(n)$, and thus the sequence

(10.3.1) $$0 \longrightarrow O(n) \longrightarrow F(n) \xrightarrow{\pi_{A'}\nabla^{A'}_A} F_A(n+1)$$

is exact, where $F_A(n+1)$ is the sheaf of germs of spinor-valued holomorphic functions of $x^{AA'}$ and $\pi_{A'}$, homogeneous of degree n+1 in $\pi_{A'}$. Sequence (10.3.1) can be completed as follows:

(10.3.2) $$\begin{array}{ccccccccc} 0 \longrightarrow O(n) \longrightarrow & F(n) & \xrightarrow{\pi_{A'}\nabla^{A'}_A} & & F_A(n+1) & \xrightarrow{\pi_{B'}\nabla^{B'A}} F(n+2) \longrightarrow 0 \\ & & \searrow & & \nearrow & \\ & & & A(n+1) & & \\ & & \nearrow & & \searrow & \\ & 0 & & & & 0 \end{array} ,$$

where the auxilliary sheaf $A(n+1)$ defined by

(10.3.3) $$A(n+1) = \text{Image}(\pi_{A'}\nabla^{A'}_A) = \text{Kernel}(\pi_{B'}\nabla^{B'A})$$

has been introduced in order to facilitate the disintegration of (10.3.2) into a pair of short exact sequences:

(10.3.4) $$0 \longrightarrow O(n) \longrightarrow F(n) \longrightarrow A(n+1) \longrightarrow 0$$

(10.3.5) $$0 \longrightarrow A(n+1) \longrightarrow F_A(n+1) \longrightarrow F(n+2) \longrightarrow 0$$

Now let us consider, as an example, the helicity 1/2 case $n = -3$. The long exact cohomology sequence associated with (10.3.4) contains the segment

$$H^0(M,A(-2)) \longrightarrow H^1(M,O(-3)) \longrightarrow H^1(M,F(-3)) \longrightarrow H^1(M,A(-2)) \quad , \tag{10.3.6}$$

and associated with (10.3.5) we have the following segments:

$$0 \longrightarrow H^0(M,A(-2)) \longrightarrow H^0(M,F_A(-2)) \tag{10.3.7}$$

$$H^0(M,F(-1)) \longrightarrow H^1(M,A(-2)) \longrightarrow H^1(M,F_A(-2)) \quad . \tag{10.3.8}$$

Now $H^0(M,F_A(-2))$ and $H^0(M,F(-1))$ both consist of global functions of $x^{AA'}$ and $\pi_{A'}$ which have negative twist in $\pi_{A'}$; consequently they must both vanish. From (10.3.7) we then obtain that $H^0(M,A(-2))$ vanishes, and from (10.3.8) we deduce that the map from $H^1(M,A(-2))$ to $H^1(M,F_A(-2))$ is injective. Gathering these facts together we deduce that the sequence

$$0 \longrightarrow H^1(M,O(-3)) \longrightarrow H^1(M,F(-3)) \longrightarrow H^1(M,F_A(-2)) \tag{10.3.9}$$

is exact. The group $H^1(M,F(-3))$ is the set of all primed spinor-valued functions $f_{A'}(x)$ defined over the region of spacetime corresponding to M, whereas $H^1(M,F_A(-2))$ is the set of all unprimed spinor-valued functions $f_A(x)$ on the same region. It is not difficult to verify that the induced map between $H^1(M,F(-3))$ and $H^1(M,F_A(-2))$ is given by

$$f_{A'}(x) \longrightarrow \nabla_A^{A'} f_{A'}(x) \quad , \tag{10.3.10}$$

for a typical element. Sequence (10.3.9) asserts that $H^1(M,O(-3))$ is precisely the <u>kernel</u> of this map, and a glance at (10.3.10) shows that the kernel consists precisely of ZRM fields of the appropriate helicity.

Thus we have established Proposition 10.2.1 again, but from a slightly different point of view. In fact, a somewhat stronger result has been obtained, namely an <u>isomorphism</u> between the twistor cohomology group of interest and the relevant set of ZRM fields. It is perhaps instructive to arrive at this result from yet another angle, using an exact sequence which codifies more directly the procedure outlined in Proposition 10.2.1. Let us denote by $\Phi_{A'}$ the subsheaf of $F_{A'}$ consisting of germs satisfying the zero rest mass equations: $\nabla^{AA'}\phi_{A'}(x,\pi) = 0$. Then, for example, the following sequence is exact:

$$0 \longrightarrow O(-3) \xrightarrow{\pi_{A'}\pi_{B'}} \Phi_{A'B'}(-1) \xrightarrow{\pi^{B'}} \Phi_{A'}(0) \longrightarrow 0 \quad . \tag{10.3.11}$$

To calculate the cohomology of the Φ sheaves, one can use the following sequences:

$$0 \longrightarrow \Phi_{A'}(n) \longrightarrow F_{A'}(n) \xrightarrow{\nabla^{A'A}} F^{A}(n) \longrightarrow 0 \tag{10.3.12}$$

$$0 \longrightarrow \Phi_{A'B'}(n) \longrightarrow F_{A'B'}(n) \xrightarrow{\nabla^{B'}_{A}} F_{AA'}(n) \xrightarrow{\nabla^{AA'}} F(n) \longrightarrow 0 \quad . \tag{10.3.13}$$

Now the long exact cohomology sequence associated with (10.3.11) contains the segment

$$H^0(M,\Phi_{A'B'}(-1)) \longrightarrow H^0(M,\Phi_{A'}(0)) \longrightarrow H^1(M,O(-3)) \longrightarrow H^1(M,\Phi_{A'B'}(-1)) \; . \tag{10.3.14}$$

Using (10.3.13) one can establish that both H^0 and H^1 vanish for $\Phi_{A'B'}(-1)$, whence we have the isomorphism

$$0 \longrightarrow H^0(M,\Phi_{A'}(0)) \longrightarrow H^1(M,O(-3)) \longrightarrow 0 \; . \tag{10.3.15}$$

And finally, an elementary calculation using (10.3.12) shows that $H^0(M,\Phi_{A'}(0))$ consists of the relevant set of zero rest mass fields.

For positive helicity ($s \geq 0$) sequence (10.3.11) can be generalized as follows:

$$0 \longrightarrow O(-2s-2) \xrightarrow{\alpha} \Phi_{A'\ldots B'C'}(-1) \xrightarrow{\beta} \Phi_{A'\ldots B'}(0) \longrightarrow 0 \quad , \tag{10.3.16}$$

where the map α is multiplication by $\pi_{A'}\ldots\pi_{B'}\pi_{C'}$, the total number of π's being $2s+1$. The map β is contraction with $\pi^{C'}$.

For <u>negative</u> helicity, the analogue of (10.3.16) is the following sequence:

$$0 \longrightarrow P(-2s-2) \longrightarrow O(-2s-2) \xrightarrow{\alpha} \Phi_{A\ldots B} \xrightarrow{\beta} \Phi_{A\ldots BC} \longrightarrow 0 \quad . \tag{10.3.17}$$

The sheaf $P(-2s-2)$ consists of polynomials in ω^A and $\pi_{A'}$, homogeneous of degree $-2s-2$. The map α is $\hat{\pi}_A\ldots\hat{\pi}_B$ ($-2s-3$ occurrences of $\hat{\pi}_A$), and β is the operator $\pi_{A'}\nabla^{A'}_{C}$.

By using (10.3.16) and (10.3.17), one can establish again the connection between twistor cohomology and ZRM fields, for all helicities. These two sequences, incidentally, are the original sequences suggested by R. Penrose in order to solve the problem of relating ZRM fields and twistor sheaf cohomology groups[(3)]. The sequence (10.3.2) is somewhat more advantageous, at least in certain respects, inasmuch as the spin-bundle sheaves F(n), etc., are easy to work with directly—this point of view has been reflected in work by M. Eastwood, R. Penrose, and R.O. Wells, Jr.. As another alternative to sequences (10.3.16) and (10.3.17) one can consider the single sequence

$$(10.3.18) \qquad 0 \longrightarrow O(-2s-2) \xrightarrow{\pi_{A'}} O_{A'}(-2s-2) \xrightarrow{\pi^{A'}} O(-2s) \longrightarrow 0$$

which can be used to derive the relevant twistor cohomology relationships for all helicities—as with sequence (10.3.2), certain complications arise on account of the appearance of potentials in the various formulae, but these complications are not particularly serious[(4)].

10.4 Remarks on the Geometry of n-Twistor Systems.

Now we shall begin to set up some of the machinery necessary for the treatment of more general categories of fields, using twistor sheaf cohomological methods. Our attitude will be to regard n-twistor systems from a somewhat more abstract point of view, treating them as higher-dimensional complex spaces with relatively little structure at first, and then introducing more and more structure piece by piece. In this way we arrive at the conclusion that certain operators (and their associated eigenvalues) are more "primitive" than others, since they can be introduced at an earlier stage in the whole process and require less structure in their initial definition.

Let S^a denote C^{m+1} $(m \geq 1)$ regarded as a complex vector space, and denote by S_a , $S^{a'}$, and $S_{a'}$ the dual space, the complex conjugate space, and the dual complex conjugate space to S^a , respectively. One can introduce the idea of a "generalized twistor" as a point in the space $(S^a , S_{a'})$. Thus by a generalized twistor one means a pair $(\omega^a , \pi_{a'})$ with $\omega^a \epsilon S^a$ and $\pi_{a'} \epsilon S_{a'}$.

The dual space to $(S^{a}, S_{a'})$ is the space $(S_{a}, S^{a'})$. If $(\sigma_{a}, \tau^{a'})$ is a dual twistor [henceforth we shall drop the adjective "generalized" when there is no opportunity for confusion] then its inner product with the twistor $(\omega^{a}, \pi_{a'})$ is defined to be $\omega^{a}\sigma_{a} + \pi_{a'}\tau^{a'}$. The complex conjugate of $(\omega^{a}, \pi_{a'})$ is the dual twistor $(\bar{\pi}_{a}, \bar{\omega}^{a'})$, and the <u>norm</u> of $(\omega^{a}, \pi_{a'})$ is defined by the inner product

$$\omega^{a}\bar{\pi}_{a} + \pi_{a'}\bar{\omega}^{a'} , \tag{10.4.1}$$

which, using a standard argument [cf. Section 2.3], can be shown to have signature (m+1 , m+1). The space $(S^{a}, S_{a'})$ is clearly C^{2m+2}, and the associated projective space is P^{2m+1}. The space C^{2m+2} is divided into three regions, denoted C_{+}^{2m+2}, C_{0}^{2m+2}, and C_{-}^{2m+2}, according as to whether the norm (10.4.1) is positive, zero, or negative; the three corresponding regions of P^{2m+1} are denoted P_{+}^{2m+1}, P_{0}^{2m+1}, and P_{-}^{2m+1}.

In the case of m = 1, we recover standard twistor space, with its usual norm of signature (2,2). Spaces of n twistors fit into the picture for <u>odd</u> values of m, with the identifications

$$\omega^{a} = \omega_{i}^{A} , \qquad \pi_{a'} = \pi_{iA'} . \tag{10.4.2}$$

The spaces with m <u>even</u> do not admit of an obvious spacetime interpretation, although it must be admitted that they do fit almost uncomfortably naturally into the general scheme.

The Grassmannian of projective m-planes in P^{2m+1} has dimension $(m+1)^{2}$. In the case m = 1 [where the 1-planes are, of course, lines] the Grassmannian can be regarded as complexified compactified Minkowski space. For general m a similar interpretation is available, and for the "finite" points of the associated $(m+1)^{2}$-dimensional "hyperspace"—to borrow a convenient term from the literature of science fiction—we can introduce a set of variables $x^{aa'}$ as coordinates. The Grassmannian of projective j-planes in P^{k} is often denoted G(j,k). The space G(1,3) can be realized as a quadric hypersurface in P^{5}. In the general case of interest here, the space G(m,2m+1) can be realized as an algebraic variety of dimension $(m+1)^{2}$ in P^{r}, where r is given by the formula

$$r = (2m+2)!/[(m+1)!]^2 \qquad (10.4.3)$$

this being the dimension of the space of skew-symmetric tensors of valence m+1 in C^{2m+2}.

In the case m=1 there is a "preferred" line P^1_I in P^3 corresponding to the vertex of scri in spacetime. The remaining points of scri correspond to lines in P^3 which meet P^3_I. The "finite" points of spacetime correspond to lines in P^3 which avoid P^1_I. Likewise, there is a preferred m-plane P^m_I in P^{2m+1}. The "finite" points of the hyperspace G(m,2m+1) correspond, by definition, to m-planes in P^{2m+1} which avoid P^m_I. "Infinity" in G(m,2m+1) consists of points which correspond to m-planes in P^{2m+1} which intersect P^m_I. The points of G(m,2m+1) can be classified by a number d which is the dimension of the intersection of the corresponding m-plane with P^m_I. For finite points we put d = -1. In the case of standard twistor space, given by m=1, there are just three classes of associated spacetime points: finite points (d = -1), non-vertex points on scri (d = 0), and the vertex of scri (d = 1). In the general case d has the range d = -1,0,...,m; the case d = m corresponding to the m-plane P^m_I itself.

A generalized twistor $(\omega^a, \pi_{a'})$ lies on the m-plane corresponding to the point $x^{aa'}$ if and only if the relation

$$\omega^a = ix^{aa'}\pi_{a'} \qquad (10.4.4)$$

holds. We can characterize generalized twistors in terms of solutions of the equation

$$(m+1)\nabla_{aa'}\xi^b = \delta^b_a\nabla_{ca'}\xi^c \qquad (10.4.5)$$

In fact, we have the following result (cf. Hughston, 1979), which is analogous to Proposition 2.4.2:

<u>10.4.6 Theorem</u>. The general solution of equation (10.4.5) is given by

$$\xi^a = \omega^a - ix^{aa'}\pi_{a'} \qquad (10.4.7)$$

where ω^a and $\pi_{a'}$ are constant.

Proof. Differentiating (10.4.5) one has

(10.4.8) $$(m+1)\nabla_{bb'}\nabla_{aa'}\xi^c = \delta^c_a\nabla_{bb'}\nabla_{da'}\xi^d \quad ,$$

which, exchanging aa' and bb' , can be written as

(10.4.9) $$(m+1)\nabla_{aa'}\nabla_{bb'}\xi^c = \delta^c_b\nabla_{aa'}\nabla_{db'}\xi^d \quad .$$

Since the ∇'s commute, the left hand sides of equations (10.4.8) and (10.4.9) are equal; therefore:

(10.4.10) $$\delta^c_a\nabla_{bb'}\nabla_{da'}\xi^d = \delta^c_b\nabla_{aa'}\nabla_{db'}\xi^d \quad .$$

Transvecting equation (10.4.10) with δ^b_c gives

(10.4.11) $$\nabla_{ab'}\nabla_{da'}\xi^d = (m+1)\nabla_{aa'}\nabla_{db'}\xi^d \quad .$$

On the other hand, if equation (10.4.10) is transvected instead with δ^a_c one obtains $(m+1)\nabla_{bb'}\nabla_{da'}\xi^d = \nabla_{ba'}\nabla_{db'}\xi^d$ which, substituting b with a , reads:

(10.4.12) $$(m+1)\nabla_{ab'}\nabla_{da'}\xi^d = \nabla_{aa'}\nabla_{db'}\xi^d \quad .$$

For $m > 0$, equations (10.4.11) and (10.4.12) together imply

(10.4.13) $$\nabla_{aa'}\nabla_{db'}\xi^d = 0 \quad ,$$

showing that $\nabla_{db'}\xi^d$ is constant; from (10.4.5) we can therefore infer that ξ^a is linear in $x^{aa'}$. Inserting the most general linear expression for ξ^a into (10.4.5), the desired result (10.4.7) follows after a short calculation[(5)]. □

The locus of a twistor $(\omega^a, \pi_{a'})$ is defined to be the set of points in the space G(m,2m+1) corresponding to the pencil of m-planes in P^{2m+1} through the point $(\omega^a, \pi_{a'})$. From (10.4.4) and (10.4.7) it follows that the locus, insofar as finite points are concerned, is the region where the associated spinor field $\xi^a(x)$ vanishes.

When m is odd, and the n-twistor relations (10.4.2) are assumed, equation (10.4.4) reads

(10.4.14) $$\omega_i^A = ix_i^{AA'j}\pi_{jA'} \quad ,$$

where we have made the identification

(10.4.15) $$x^{aa'} = x_i^{AA'j}$$

for the coordinates of finite points in G(m,2m+1). Complex Minkowski space is embedded naturally as a linear subspace in this space, given by

(10.4.16) $$x_i^{AA'j} = x^{AA'}\delta_i^j \quad .$$

We shall take the attitude that fields on spacetime can be regarded as restrictions, down to spacetime, of "hyperspace" fields. Thus, a field $\emptyset(x^{AA'})$ on spacetime should be thought of as the restriction ρ_x of some hyperspace field $\Phi(x_i^{AA'j})$ according to the following scheme:

(10.4.17) $$\emptyset(x^{AA'}) = \rho_x\Phi(x_i^{AA'j}) = \Phi(x^{AA'}\delta_i^j) \quad .$$

By requiring that the hyperspace field $\Phi(x_i^{AA'j})$ exhibit suitable properties, conditions can be imposed on the spacetime field $\emptyset(x)$ projected from it.

10.5 Massive Fields Revisited.

The contour formulae introduced for massive fields in Section 5.2 can be reinterpreted in an interesting way in the light of the remarks made in Section 10.4.

In Section 5.2 we were concerned with contour integral formulae of the form

(10.5.1) $$\emptyset^{A\ldots j\ldots}_{A'\ldots i\ldots}(x) = \oint \rho_x \hat{\pi}^{Aj}\ldots\pi_{A'i}f(Z_i^\alpha)\Delta\pi \quad .$$

Using the notation of Section 10.4, formula (10.5.1) can be rewritten as:

(10.5.2) $$\emptyset_{a\ldots a'\ldots}(x) = \oint \rho_x\hat{\pi}_a\ldots\pi_{a'}\ldots f(\omega^a,\pi_{a'})\Delta\pi \quad ,$$

where, in addition to (10.4.2), we use the notation $\hat{\pi}_a = -\partial/\partial\omega^a$, along with

(10.5.3) $$\rho_x f(\omega^a,\pi_{a'}) = \rho_x f(\omega_i^A,\pi_{A'i}) = f(ix^{AA'}\pi_{A'},\pi_{A'}) \quad .$$

Now it turns out that formula (10.5.2) can be evaluated in two steps. First we evaluate the twistor function $f(\omega^{a},\pi_{a'})$ on hyperspace (using a suitable set of π-coefficients, as will be described below). Then, we take a number of derivatives of the hyperfield, using the operator $\nabla^{i}_{A'Aj} = \partial/\partial x^{AA'j}_{i}$, and restrict the result down to spacetime. This gives us the field $\phi_{a\ldots a'\ldots}(x)$.

Suppose the number of π-coefficients in (10.5.2) is denoted π, and the number of $\hat{\pi}$-coefficients is denoted $\hat{\pi}$. The coefficient structure for the hyperspace evaluation is determined as follows. If $\hat{\pi}-\pi$ is positive, we use $\hat{\pi}-\pi$ coefficients of the $\hat{\pi}$-type. If $\pi-\hat{\pi}$ is positive, on the other hand, we use $\pi-\hat{\pi}$ coefficients of the π-type.

The number of derivatives we take before restricting the hyperfield down to spacetime is simply the absolute value $|\hat{\pi}-\pi|$.

It should be observed that when we examine hyperfields, we need only consider fields whose indices are all of the same type. Let us denote by $\rho_{\underline{x}}$ the restriction down to the hyperspace point $x^{aa'}$. Then the following three contour integral formulas are of interest to us:

(10.5.4) $$\Phi_{a\ldots b}(\underline{x}) = \oint \rho_{\underline{x}}\hat{\pi}_{a}\ldots\hat{\pi}_{b}f(\omega^{a},\pi_{a'})\Delta\pi$$

(10.5.5) $$\Phi(\underline{x}) = \oint \rho_{\underline{x}}f(\omega^{a},\pi_{a'})\Delta\pi$$

(10.5.6) $$\Phi_{a'\ldots b'}(\underline{x}) = \oint \rho_{\underline{x}}\pi_{a'}\ldots\pi_{b'}f(\omega^{a},\pi_{a'})\Delta\pi$$

By taking derivatives, and using the identity

(10.5.7) $$i\nabla_{aa'}\rho_{\underline{x}} = \rho_{\underline{x}}\hat{\pi}_{a}\pi_{a'}$$

which is valid for functions of ω^{a} and $\pi_{a'}$, we can recover general fields of the form (10.5.2).

10.6 Towards the Cohomology of n-Twistor Systems.

The fields defined in formulae (10.5.4), (10.5.5), and (10.5.6) exhibit a rather curious feature; they satisfy the following set of field equations:

(10.6.1) $$\nabla_{a'[a}\Phi_{b]\ldots c} = 0 \quad ,$$

(10.6.2) $$\nabla_{a'[a}\nabla_{b]b'}\Phi = 0 \quad ,$$

(10.6.3) $$\nabla_{a[a'}\Phi_{b']\ldots c'} = 0 \quad .$$

These relations generalize the ZRM equations in a natural way, and in the case m = 1 they reduce to the ZRM equations. Note that in the case of a scalar field Φ we have a second order equation (analogous to the wave equation), whereas in the other case we have first order equations. However, as was pointed out by M. Eastwood, equations (10.6.1) and (10.6.3) imply the second order equations

(10.6.4) $$\nabla_{p'[p}\nabla_{q]q'}\Phi_{a\ldots b} = 0 \quad ,$$

(10.6.5) $$\nabla_{p'[p}\nabla_{q]q'}\Phi_{a'\ldots b'} = 0 \quad ,$$

as one might expect by analogy with the case m = 1.

We can classify the Φ-fields with a half-integer r. When r is positive we have a field with 2r primed indices, and when r is negative we have a field with -2r unprimed indices. When r = 0 we have a scalar field. For lack of better terminology, we shall refer to r as the "hyperhelicity".

It turns out, rather remarkably, that solutions of the free field equations (10.6.1), (10.6.2), and (10.6.3) can be described as elements of certain cohomology groups. The relevant groups can be described as follows. Let M be a region of P^{2m+1} swept out by a set of m-planes corresponding to the points in a region $\tilde{M}$ of hyperspace. Recall that if we are dealing with n-twistor systems, then m = 2n-1 . Denote by O(q) the sheaf of germs of holomorphic functions on M, twisted by q. Then we have:

<u>10.6.6 Proposition</u>. Elements of the group

(10.6.7) $$H^m(\mathbf{M}, O(-2r-m-1))$$

correspond to solutions of the hyperspace free field equations (10.6.1), (10.6.2), and (10.6.3) for hyperhelicity r, over the domain $\tilde{M}$.

Proof. We shall establish the result explicitly for the case m = 2, with r = 1/2. Unfortunately, this is not one of the cases which readily admits of a spacetime interpretation (for which m must be odd); but this is not a serious drawback, since the more general cases can be inferred directly from the method that will be outlined here. Our proof will follow rather closely the material in Section 9.3.

We are concerned with the group $H^2(M^5, O(-4))$, where M^5 is a region of P^5 swept out by a set of 2-planes. Let f_{ijk} be a representative cocycle. Restricting down to the 2-plane $x^{aa'}$ we have

(10.6.8) $$\rho_{\underline{x}} \pi_{a'} \pi_{b'} \overset{-4}{f}_{ijk} = \rho_{[i} \overset{-2}{f}_{jk]a'b'}(\underline{x},\pi)$$

since H^2 is trivial for twist greater than -3, on P^2. Skewing with $\pi_{c'}$, we get

(10.6.9) $$\rho_{[i} \overset{-2}{f}_{jk]a'[b'} \pi_{c']} = 0$$

from which we deduce the existence of an $\overset{-1}{f}_{ka'b'c'}$ such that

(10.6.10) $$\overset{-2}{f}_{jka'[b'} \pi_{c']} = \rho_{[j} \overset{-1}{f}_{k]a'b'c'} \quad .$$

Skewing with $\pi_{d'}$ we get

(10.6.11) $$\rho_{[j} \overset{-1}{f}_{k]a'[b'c'} \pi_{d']} = 0 \quad ,$$

whence we obtain:

(10.6.12) $$\overset{-1}{f}_{ka'[b'c'} \pi_{d']} = \rho_k \overset{0}{f}_{a'b'c'd'} \quad .$$

Now since $\overset{0}{f}_{a'b'c'd'}$ is global and has twist zero, it must be a function of $x^{aa'}$ alone; thus we obtain our field

(10.6.13) $$\Phi_{a'}(x^{aa'}) = \overset{0}{f}_{a'b'c'd'} \varepsilon^{b'c'd'} \quad .$$

To show that $\Phi_{a'}$ satisfies the field equation

(10.6.14) $$\nabla_{b[b'} \Phi_{a']} = 0$$

is a somewhat more insidious operation. We proceed as follows:

Let D be any operator which annihilates the expression $\rho_x \pi_{a'} \pi_{b'} f_{ijk}$. In our case D is defined by

$$\text{(10.6.15)} \qquad D: \rho_x \pi_{a'} \pi_{b'} f_{ijk} \longrightarrow \nabla_{e[e'} \pi_{a']} \pi_{b'} \rho_x f_{ijk}$$

which vanishes; but the results which follow are independent of the specific choice of D. From (10.6.8) we obtain

$$\text{(10.6.16)} \qquad \rho_{[i} D\overset{-2}{f}_{jk]a'b'} = 0$$

whence we have

$$\text{(10.6.17)} \qquad D\overset{-2}{f}_{jka'b'} = \rho_{[j} \overset{-2}{f}_{k]a'b'}$$

for some $\overset{-2}{f}_{ka'b'}$. Applying $\pi_{c'}$ and skewing, we get

$$\text{(10.6.18)} \qquad D\overset{-2}{f}_{jka'[b'} \pi_{c']} = \rho_{[j} \overset{-2}{f}_{k]a'[b'} \pi_{c']} \quad .$$

Now suppose we take (10.6.10), and hit it with D. Then we obtain

$$\text{(10.6.19)} \qquad D\overset{-2}{f}_{jka'[b'} \pi_{c']} = \rho_{[j} D\overset{-1}{f}_{k]a'b'c'} \quad .$$

Since the right hand sides of (10.6.18) and (10.6.19) are equal, we have

$$\text{(10.6.20)} \qquad \rho_{[j} \overset{-2}{f}_{k]a'[b'} \pi_{c']} = \rho_{[j} D\overset{-1}{f}_{k]a'b'c'} \quad .$$

which asserts that the cochain

$$\text{(10.6.21)} \qquad \overset{-2}{f}_{ka'[b'} \pi_{c']} - D\overset{-1}{f}_{ka'b'c'}$$

is <u>global</u>. Since the twist is negative, it follows that (10.6.21) must <u>vanish</u>, whence we deduce, skewing $\pi_{d'}$, that

$$\text{(10.6.22)} \qquad D\overset{-1}{f}_{ka'[b'c'} \pi_{d']} = 0 \quad ,$$

from which we get $D\Phi_{a'} = 0$, using (10.6.12) and (10.6.13). However, it is not difficult to verify that if we trace the action of D through the various formulae above then (10.6.15) implies

(10.6.23) $\quad D: \quad \Phi_{a'} \longrightarrow \nabla_{e[e'}\Phi_{a']}$

showing that $\Phi_{a'}$ satisfies the field equations, as desired[6]. □

It would be nice to sharpen up Proposition 10.6.6 a bit, so as to specify precisely for what sort of domains an actual isomorphism is obtained between H^m and the relevant set of hyperfields. A reasonable candidate for M is the space P_+^{4n-1} (or possibly its closure), which in twistor terms is the space $Z_i^{\alpha}\bar{Z}_{\alpha}^{i} > 0$. The (2n-1)-planes lying entirely within P_+^{4n-1} include, as a subset, a four-dimensional family of planes corresponding to the future tube CM^+ in complex Minkowski space. The "Minkowskian" (2n-1)-planes are obtained by looking at n-twistors which are of the form $Z_j^{\alpha} = (ix^{AA'}\pi_{A'j}, \pi_{A'j})$. For each fixed value of $x^{AA'}$ we obtain a (2n-1)-plane by varying $\pi_{A'j}$. Note that $x^{AA'}$ is in CM^+ if and only if $Z_i^{\alpha}\bar{Z}_{\alpha}^{i} = i(x^{AA'} - \bar{x}^{AA'})\pi_{A'j}\bar{\pi}_A^j$ is strictly positive for all values of $\pi_{A'i}$.

Before an element of H^{2n-1} can be used to represent a particle state, it must be put into an eigenstate of a suitable set of observables, as described in Chapters 5, 6, 7, 8. One particular observable—namely, the "hyperhelicity"—is already implicit in the construction, since it enters into the twist of the sheaf. In the case of 3-twistor functions, the hyperhelicity is $-3B$, with B the baryon number. If the cocycle representing some element of H^{2n-1} is placed into an eigenstate of a suitably "complete" set of observables, then the associated Φ-hyperfield will restrict down—after an appropriate number of derivatives have been taken—to an essentially unique Minkowski space wave function. The restrictions of higher derivatives of the Φ-hyperfield down to Minkowski space contain no essential new information, but do include spacetime derivatives of the basic field already obtained.

Chapter 10, Notes

1. The Kerr theorem is discussed in Penrose (1967), to which the reader is referred for further details.

2. For an account of the 27 lines on the cubic surface in P^3 see, for example,

Mumford (1976). In the case of twistor theory the quadric surface in P^3 is also of special interest. Associated with such a surface is a solution of the twistor equation of valence two:

$$\nabla^{A(A'}\xi^{B'C')} = 0 \quad .$$

This differential equation is rather curious inasmuch as it admits non-trivial solutions in curved spacetime. In particular, the Kerr solution of Einstein's equations possesses a solution of the valence two twistor equation. For discussion related to this matter see Carter (1968), Walker and Penrose (1970), Hughston, Penrose, Sommers, and Walker (1972), Hughston and Sommers (1973a and 1973b), and Sommers (1973).

3. These results, which were first described in Twistor Newsletter in 1977, appear in Penrose (1979).

4. For more extensive accounts of the material described in this section see, for example, Jozsa (1976), Pratt (1977), Moore (1978), Burnett-Stuart (1978), Weber (1978), Ward (1979), and Wells (1979). There are also to be found numerous Twistor Newsletter articles on the subject.

5. It is also interesting to note that the geodesic shearfree condition generalizes in an interesting way to the cases for which m is greater than one. One obtains the following formula:

$$(\xi^{[a'}\nabla^{b']b}\xi^{[c'})\xi^{d']} = 0 \quad .$$

As will be described elsewhere, there exists an analogue of the Kerr theorem which allows one to solve this equation using complex analytic methods.

6. I am indebted to M. Eastwood for a number of illuminating discussions in connection with the material of Section 10.6.

REFERENCES

Atiyah, M.F., 1976, "Singularities of Functions", Bulletin of the Institute of Mathematics and its Applications, Vol. 12, No. 7, 203.

Atiyah, M.F., N.J. Hitchin, and I.M. Singer, 1977, "Deformations of Instantons", Proc. Nat. Acad. Sci. USA 74, 2662.

Atiyah, M.F., and R.S. Ward, 1977, "Instantons and Algebraic Geometry", Comm. Math. Phys. 55, 117.

Beck and Daniel, 1968, Zeit. Phys. 216, 229.

Bergkvis, K.E., 1972, Nuclear Physics B39, 317.

Borer, K., et al, 1969, Phys. Lett. 29B, 614.

Burnett-Stuart, G., 1978, "Deformed Twistor Spaces", M. Sc. thesis, University of Oxford.

Carter, B., 1968, Comm. Math. Phys. 10, 280.

Chern, S.S., 1967, Complex Manifolds without Potential Theory (Van Nostrand).

Chomsky, N., 1965, Aspects of the Theory of Syntax (M.I.T. Press).

Clark, A.R., et al, 1974, Phys. Rev. D9, 533.

Dalitz, R.H., 1966, in High Energy Physics, edited by C. DeWitt and M. Jacob, (Gordon and Breach).

Feld, B.T., 1969, Models of Elementary Particles (Blaisdell Publishing Company).

Feynman, R.P., 1972, Photon-Hadron Interactions (Benjamin).

Feynman, R.P., and M. Gell-Mann, 1958, Phys. Rev. 109, 193.

Feynman, R.P., M. Kislinger, and F. Ravndal, 1971, Phys. Rev. D3, 11, 2706.

Gell-Mann, M., and Y. Ne'eman, 1964, The Eightfold Way (Benjamin).

Godement, R., 1973, Théorie des Faisceaux (Hermann, Paris).

Goldhaber, M., L. Grodzins, and A.W. Sunyar, 1958, Phys. Rev. 109, 1015.

Griffiths, P., and J. Adams, 1974, Topics in Algebraic and Analytic Geometry (Princeton University Press).

Gunning, R.C., 1966, Lectures on Riemann Surfaces (Princeton University Press).

Gunning, R.C., and H. Rossi, 1965, Analytic Functions of Several Complex Variables (Prentice-Hall).

Harari, H., 1969, Phys. Rev. Letts., 22, 562.

Harris, A., 1975, M. Sc. thesis, University of Oxford.

Hartshorne, R., 1977, Algebraic Geometry, Springer Graduate Texts in Mathematics.

Hartshorne, R., 1978, "Stable Vector Bundles and Instantons", Comm. Math. Phys. 59, 1-15.

Hawking, S.W., and G.F.R. Ellis, 1973, The Large Scale Structure of Spacetime (Cambridge University Press).

Hodges, A.P., 1975, Ph.D. thesis, Birkbeck College, University of London.

Huggett, S.A., 1976, M.Sc. thesis, University of Oxford.

Hughston, L.P., 1979, "Some New Contour Integral Formulae", in Complex Manifold Techniques in Theoretical Physics, edited by D. Lerner and P. Sommers (Pitman Publishing Co., London).

Hughston, L.P., and M. Sheppard, 1979, "On the Magnetic Moments of Hadrons", Reports Math. Phys. (to be published).

Hughston, L.P., R. Penrose, P. Sommers, and M. Walker, 1972, "On a Quadratic First Integral for the Charged Particle Orbits in the Charged Kerr Solution", Comm. Math. Phys., 27, 303.

Hughston, L.P., and P. Sommers, 1973b, "The Symmetries of Kerr Black Holes", Comm. Math. Phys. 33, 129.

Hughston, L.P., and P. Sommers, 1973a, "Spacetime with Killing Tensors", Comm. Math. Phys. 32, 147.

Jacob, M., 1974, editor, Dual Theory (North-Holland).

Jozsa, R., 1976, M.Sc. thesis, University of Oxford.

Kerr, R.P., 1963, Phys. Rev. Letts. 11, 237.

Kikkawa, K., et al, 1969, Phys. Rev. 184, 1701.

Kodaira, K., and D.C. Spencer, 1958, Ann. Math., 67, 328, 403.

Kodaira, K., and D.C. Spencer, 1960, Ann. Math., 71, 43.

Marshak, R.E., and E.C.G. Sudarshan, 1958, Phys. Rev. 109, 1860.

Matsuoka, T., et al, 1969, Prog. Theor. Phys. 42, 56.

Morrow, J., and K. Kodaira, 1971, Complex Manifolds (Holt, Rienhard, and Winston).

Moore, R., 1978, M.Sc. thesis, University of Oxford.

Mumford, D., 1976, Algebraic Geometry I: Complex Projective Varieties, Springer-Verlag.

Neville, D., 1969, Phys. Rev. Letts. 22, 494.

Newman, E.T., and J. Winicour, 1974, J. Math. Phys. 15, 113.

Particle Data Group, 1978, "Review of Particle Properties", Phys. Letters, Vol. 75B, No. 1.

Penrose, R., 1963, "Asymptotic Properties of Fields and Spacetimes", Phys. Rev. Letts. 10, 66.

Penrose, R., 1965a, "Zero-Rest-Mass Fields Including Gravitation: Asymptotic Behavior", Proc. Roy. Soc. A284, 159.

Penrose, R., 1965b, "Conformal Treatment of Infinity", in Relativity, Groups, and Topology, edited by B. DeWitt and C.M. DeWitt (Gordon and Breach).

Penrose, R., 1967, "Twistor Algebra", J. Math. Phys., Vol. 8, No. 2, 345.

Penrose, R., 1968a, "The Structure of Spacetime" in Battelle Renontres 1967, edited by C.M. DeWitt and J.A. Wheeler (Benjamin).

Penrose, R., 1968b, "Twistor Quantization and Curved Space-Time", Int. J. Theor. Phys., Vol. 1, No. 1, 61.

Penrose, R., 1969, J. Math. Phys. 10, 38.

Penrose, R., 1971a, in Combinatorial Mathematics and its Applications, edited by D.J.A. Welsh (Academic Press).

Pernose, R., 1971b, in Quantum Theory and Beyond, edited by T. Bastin (Cambridge University Press).

Penrose, R., 1972, "Twistor Theory: An Approach to the Quantization of Fields and Spacetime", Phys. Reports, Vol. 6C, No. 4.

Penrose, R., 1975a, "Twistor Theory, Its Aims and Achievements", in Quantum Gravity: An Oxford Symposium, edited by C.J. Isham, R. Penrose, and D.W. Sciama (Clarendon Press, Oxford).

Penrose, R., 1975b, "Twistors and Particles", in Quantum Theory and the Structure of Time and Space, edited by L. Castell, M. Drieschner, and C.F. von Weiszacker (Munchen: Verlag).

Penrose, R., 1976, "Curved Twistor Theory and Non-Linear Gravitons", Gen. Rel. and Grav., Vol 7, No. 1, 31.

Penrose, R., 1977, "The Twistor Programme", Reports Math. Phys. 12, 65.

Penrose, R., 1979, in Complex Manifold Techniques in Theoretical Physics, edited by D. Lerner and P. Sommers (Pitman Publishing Co., London).

Penrose, R., G.A.J. Sparling, and Tsou S.T., 1978, "Extended Regge Trajectories", J. Phys. A: Math. Gen., Vol. 11, No. 9, L231.

Perjés, Z., 1975, "Twistor Variables of Relativistic Mechanics", Phys. Rev. D 11, 2031.

Perjés, Z., 1977, "Perspectives of Penrose Theory in Particle Physics", Reports Math Phys. 12, 193.

Perjés, Z., and G.A.J. Sparling, 1976, "Evidence for the Twistor Structure of Hadrons", Hungarian Academy of Sciences preprint (Budapest).

Perl, M., et al, 1975, Phys. Rev. Letts. 35, 1489.

Perl, M., et al, 1976, Phys. Letts. 63B, 466.

Pirani, F.A.E., 1965, in Lectures on General Relativity, A. Trautman, F.A.E. Pirani, and H. Bondi, 1964 Brandeis Summer Institute on Theoretical Physics (Prentice-Hall).

Popovich, A., 1978, M.Sc. thesis, University of Oxford.

Pratt, S., 1977, M.Sc. thesis, University of Oxford.

Reines, F., and C.L. Cowan, 1953, Phys. Rev., 92, 830.

Robinson, I., 1961, "Null Electromagnetic Fields", J. Math. Phys., 2, 290.

Rosner, J., 1969, Phys. Rev. Letts. 22, 689.

Ryman, A., 1975, D. Phil. thesis, University of Oxford.

Serre, J.P., 1955, "Un Théorème de Dualité", Comm. Math. Helv. 29, 9.

Serre, J.P., 1956, "Géométrie Algébrique et Géométrie Analytique" (GAGA), Ann. Inst. Fourier 6, 1.

Sommers, P., 1973, "Killing Tensors and Type {2,2} Spacetimes", Ph.D. thesis, University of Texas at Austin.

Sparling, G.A.J., 1974, Ph.D. thesis, Birkbeck College, University of London.

Sparling, G.A.J., 1975, "Homology and Twistor Theory", in Quantum Gravity: An Oxford Symposium, edited by C.J. Isham, R. Penrose, and D.W. Sciama (Clarendon Press, Oxford).

Sparling, G.A.J., 1976, "Predictions of a Twistor Model for Leptons", University of Pittsburgh Physics Department Preprint.

Shafarevich, I.R., 1977, Basic Algebraic Geometry (Springer-Verlag).

Tod, K.P., 1975, "Massive Particles with Spin in General Relativity and Twistor Theory", D. Phil. thesis, University of Oxford.

Tod, K.P., 1977, "Some Symplectic Forms arising in Twistor Theory", Reports Math. Phys. 11, 339.

Tod, K.P., and Z. Perjés, 1976, Gen. Rel. and Grav., Vol. 7, No. 11, 903.

Walker, M., and R. Penrose, 1970, Comm. Math. Phys. 18, 265.

Ward, R.S., 1977a, "Curved Twistor Spaces", D. Phil. thesis, University of Oxford.

Ward, R.S., 1977b, Phys. Lett. 61A, 81.

Ward, R.S., 1979, in Complex Manifold Techniques in Theoretical Physics, edited by D. Lerner and P. Sommers (Pitman Publishing Co., London).

Weber, W., 1978, M.Sc. thesis, University of Oxford.

Wells, R.O., 1979, in Complex Manifold Techniques in Theoretical Physics, edited by D. Lerner and P. Sommers (Pitman Publishing Co., London).

Weinberg, S., 1967, Phys. Rev. Letts. 19, 1264.

Weinberg, S., 1971, Phys. Rev. Letts. 28, 1688.

Woodhouse, N.M.J., 1976, "Twistor Theory and Geometric Quantization", in Group Theoretical Methods in Physics, Lecture Notes on Physics, Vol. 50, (Springer).

INDEX

A new journal

Zeitschrift für Phisik C

Particles and Fields

ISSN 0170-9739 Title No. 288

Editors in Chief: G. Kramer, Hamburg; H. Satz, Bielefeld

Editors: K. Fujikawa, Tokyo; K. Gottfried, Cornell; K. Kajantie, Helsinki; A. Krzywicki, Orsay; P. Landshoff, Cambridge; J. J. Sakurai, UCLA; P. Söding, DESY; B. Stech, Heidelberg; J. Steinberger, CERN

Zeitschrift für Physik appears in three parts:

A: Atoms and Nuclei
B: Condensed Matter and Quanta
C: Particles and Fields

Each part may be ordered separately.
Coordinating editor for Zeitschrift für Physik, Parts A, B and C, is O. Haxel, Heidelberg.

Zeitschrift für Physik C – Particles and Fields is devoted to the experimental and theoretical investigation of elementary particles. In view of the steadily growing interplay of theory and experiment in this field, particular emphasis is given to a clear and complete presentation of research.
The topics covered include: strong, electromagnetic, and weak interactions of elementary particles, the constituent structure of elementary particles, interaction and classification of constituents, and symmetry and unification schemes of different interactions.

Language of publication is English.

Subscription information and sample copies upon request.

Springer-Verlag
Berlin
Heidelberg
New York

Selected Issues from

Lecture Notes in Mathematics

Vol. 561: Function Theoretic Methods for Partial Differential Equations. Darmstadt 1976. Proceedings. Edited by V. E. Meister, N. Weck and W. L. Wendland. XVIII, 520 pages. 1976.

Vol. 564: Ordinary and Partial Differential Equations, Dundee 1976. Proceedings. Edited by W. N. Everitt and B. D. Sleeman. XVIII, 551 pages. 1976.

Vol. 565: Turbulence and Navier Stokes Equations. Proceedings 1975. Edited by R. Temam. IX, 194 pages. 1976.

Vol. 566: Empirical Distributions and Processes. Oberwolfach 1976. Proceedings. Edited by P. Gaenssler and P. Révész. VII, 146 pages. 1976.

Vol. 570: Differential Geometrical Methods in Mathematical Physics, Bonn 1975. Proceedings. Edited by K. Bleuler and A. Reetz. VIII, 576 pages. 1977.

Vol. 572: Sparse Matrix Techniques, Copenhagen 1976. Edited by V. A. Barker. V, 184 pages. 1977.

Vol. 579: Combinatoire et Représentation du Groupe Symétrique, Strasbourg 1976. Proceedings 1976. Edité par D. Foata. IV, 339 pages. 1977.

Vol. 587: Non-Commutative Harmonic Analysis. Proceedings 1976. Edited by J. Carmona and M. Vergne. IV, 240 pages. 1977.

Vol. 592: D. Voigt, Induzierte Darstellungen in der Theorie der endlichen, algebraischen Gruppen. V, 413 Seiten. 1977.

Vol. 594: Singular Perturbations and Boundary Layer Theory, Lyon 1976. Edited by C. M. Brauner, B. Gay, and J. Mathieu. VIII, 539 pages. 1977.

Vol. 596: K. Deimling, Ordinary Differential Equations in Banach Spaces. VI, 137 pages. 1977.

Vol. 605: Sario et al., Classification Theory of Riemannian Manifolds. XX, 498 pages. 1977.

Vol. 606: Mathematical Aspects of Finite Element Methods. Proceedings 1975. Edited by I. Galligani and E. Magenes. VI, 362 pages. 1977.

Vol. 607: M. Métivier, Reelle und Vektorwertige Quasimartingale und die Theorie der Stochastischen Integration. X, 310 Seiten. 1977.

Vol. 615: Turbulence Seminar, Proceedings 1976/77. Edited by P. Bernard and T. Ratiu. VI, 155 pages. 1977.

Vol. 618: I. I. Hirschman, Jr. and D. E. Hughes, Extreme Eigen Values of Toeplitz Operators. VI, 145 pages. 1977.

Vol. 623: I. Erdelyi and R. Lange, Spectral Decompositions on Banach Spaces. VIII, 122 pages. 1977.

Vol. 628: H. J. Baues, Obstruction Theory on the Homotopy Classification of Maps. XII, 387 pages. 1977.

Vol. 629: W. A. Coppel, Dichotomies in Stability Theory. VI, 98 pages. 1978.

Vol. 630: Numerical Analysis, Proceedings, Biennial Conference, Dundee 1977. Edited by G. A. Watson. XII, 199 pages. 1978.

Vol. 636: Journées de Statistique des Processus Stochastiques, Grenoble 1977, Proceedings. Edité par Didier Dacunha-Castelle et Bernard Van Cutsem. VII, 202 pages. 1978.

Vol. 638: P. Shanahan, The Atiyah-Singer Index Theorem, An Introduction. V, 224 pages. 1978.

Vol. 648: Nonlinear Partial Differential Equations and Applications, Proceedings, Indiana 1976–1977. Edited by J. M. Chadam. VI, 206 pages. 1978.

Vol. 650: C*-Algebras and Applications to Physics. Proceedings 1977. Edited by R. V. Kadison. V, 192 pages. 1978.

Vol. 656: Probability Theory on Vector Spaces. Proceedings, 1977. Edited by A. Weron. VIII, 274 pages. 1978.

Vol. 662: Akin, The Metric Theory of Banach Manifolds. XIX, 306 pages. 1978.

Vol. 665: Journées d'Analyse Non Linéaire. Proceedings, 1977. Edité par P. Bénilan et J. Robert. VIII, 256 pages. 1978.

Vol. 667: J. Gilewicz, Approximants de Padé. XIV, 511 pages. 1978.

Vol. 668: The Structure of Attractors in Dynamical Systems. Proceedings, 1977. Edited by J. C. Martin, N. G. Markley and W. Perrizo. VI, 264 pages. 1978.

Vol. 675: J. Galambos and S. Kotz, Characterizations of Probability Distributions. VIII, 169 pages. 1978.

Vol. 676: Differential Geometrical Methods in Mathematical Physics II, Proceedings, 1977. Edited by K. Bleuler, H. R. Petry and A. Reetz. VI, 626 pages. 1978.

Vol. 678: D. Dacunha-Castelle, H. Heyer et B. Roynette. Ecole d'Eté de Probabilités de Saint-Flour. VII-1977. Edité par P. L. Hennequin. IX, 379 pages. 1978.

Vol. 679: Numerical Treatment of Differential Equations in Applications, Proceedings, 1977. Edited by R. Ansorge and W. Törnig. IX, 163 pages. 1978.

Vol. 681: Séminaire de Théorie du Potentiel Paris, No. 3, Directeurs: M. Brelot, G. Choquet et J. Deny. Rédacteurs: F. Hirsch et G. Mokobodzki. VII, 294 pages. 1978.

Vol. 682: G. D. James, The Representation Theory of the Symmetric Groups. V, 156 pages. 1978.

Vol. 684: E. E. Rosinger, Distributions and Nonlinear Partial Differential Equations. XI, 146 pages. 1978.

Vol. 690: W. J. J. Rey, Robust Statistical Methods. VI, 128 pages. 1978.

Vol. 691: G. Viennot, Algèbres de Lie Libres et Monoïdes Libres. III, 124 pages. 1978.

Vol. 693: Hilbert Space Operators, Proceedings, 1977. Edited by J. M. Bachar Jr. and D. W. Hadwin. VIII, 184 pages. 1978.

Vol. 696: P. J. Feinsilver, Special Functions, Probability Semigroups, and Hamiltonian Flows. VI, 112 pages. 1978.

Vol. 702: Yuri N. Bibikov, Local Theory of Nonlinear Analytic Ordinary Differential Equations. IX, 147 pages. 1979.

Vol. 704: Computing Methods in Applied Sciences and Engineering, 1977, I. Proceedings, 1977. Edited by R. Glowinski and J. L. Lions. VI, 391 pages. 1979.

Vol. 710: Séminaire Bourbaki vol. 1977/78, Exposés 507–524. IV, 328 pages. 1979.

Vol. 711: Asymptotic Analysis. Edited by F. Verhulst. V, 240 pages. 1979.

Lecture Notes in Physics

Vol. 68: Y. V. Venkatesh, Energy Methods in Time-Varying System Stability and Instability Analyses. XII, 256 pages. 1977.

Vol. 69: K. Rohlfs, Lectures on Density Wave Theory. VI, 184 pages. 1977.

Vol. 70: Wave Propagation and Underwater Acoustics. Edited by J. Keller and J. Papadakis. VIII. 287 pages. 1977.

Vol. 71: Problems of Stellar Convection. Proceedings 1976. Edited by E. A. Spiegel and J. P. Zahn. VIII, 363 pages. 1977.

Vol. 72: Les instabilités hydrodynamiques en convection libre forcée et mixte. Edité par J. C. Legros et J. K. Platten. X, 202 pages. 1978.

Vol. 73: Invariant Wave Equations. Proceedings 1977. Edited by G. Velo and A. S. Wightman. VI, 416 pages. 1978.

Vol. 74: P. Collet and J.-P. Eckmann, A Renormalization Group Analysis of the Hierarchical Model in Statistical Mechanics. IV, 199 pages. 1978.

Vol. 75: Structure and Mechanisms of Turbulence I. Proceedings 1977. Edited by H. Fiedler. XX, 295 pages. 1978.

Vol. 76: Structure and Mechanisms of Turbulence II. Proceedings 1977. Edited by H. Fiedler. XX, 406 pages. 1978.

Vol. 77: Topics in Quantum Field Theory and Gauge Theories. Proceedings, Salamanca 1977. Edited by J. A. de Azcárraga. X, 378 pages 1978.

Vol. 78: Böhm, The Rigged Hilbert Space and Quantum Mechanics. IX, 70 pages. 1978.

Vol. 79: Group Theoretical Methods in Physics. Proceedings, 1977. Edited by P. Kramer and A. Rieckers. XVIII, 546 pages. 1978.

Vol. 80: Mathematical Problems in Theoretical Physics. Proceedings, 1977. Edited by G. Dell'Antonio, S. Doplicher and G. Jona-Lasinio. VI, 438 pages. 1978.

Vol. 81: MacGregor, The Nature of the Elementary Particle. XXII, 482 pages. 1978.

Vol. 82: Few Body Systems and Nuclear Forces I. Proceedings, 1978. Edited by H. Zingl, M. Haftel and H. Zankel. XIX, 442 pages. 1978.

Vol. 83: Experimental Methods in Heavy Ion Physics. Edited by K. Bethge. V, 251 pages. 1978.

Vol. 84: Stochastic Processes in Nonequilibrium Systems, Proceedings, 1978. Edited by L. Garrido, P. Seglar and P. J. Shepherd. XI, 355 pages. 1978

Vol. 85: Applied Inverse Problems. Edited by P. C. Sabatier. V, 425 pages. 1978.

Vol. 86: Few Body Systems and Electromagnetic Interaction. Proceedings 1978. Edited by C. Ciofi degli Atti and E. De Sanctis. VI, 352 pages. 1978.

Vol. 87: Few Body Systems and Nuclear Forces II, Proceedings, 1978. Edited by H. Zingl, M. Haftel, and H. Zankel. X, 545 pages. 1978.

Vol. 88: K. Hutter and A. A. F. van de Ven, Field Matter Interactions in Thermoelastic Solids. VIII, 231 pages. 1978.

Vol. 89: Microscopic Optical Potentials, Proceedings, 1978. Edited by H. V. von Geramb. XI, 481 pages. 1979.

Vol. 90: Sixth International Conference on Numerical Methods in Fluid Dynamics. Proceedings, 1978. Edited by H. Cabannes, M. Holt and V. Rusanov. VIII, 620 pages. 1979.

Vol. 91: Computing Methods in Applied Sciences and Engineering, 1977, II. Proceedings, 1977. Edited by R. Glowinski and J. L. Lions. VI, 359 pages. 1979.

Vol. 92: Nuclear Interactions. Proceedings, 1978. Edited by B. A. Robson. XXIV, 507 pages. 1979.

Vol. 93: Stochastic Behavior in Classical and Quantum Hamiltonian Systems. Proceedings, 1977. Edited by G. Casati and J. Ford. VI, 375 pages. 1979.

Vol. 94: Group Theoretical Methods in Physics. Proceedings, 1978. Edited by W. Beiglböck, A. Böhm and E. Takasugi. XIII, 540 pages. 1979.

Vol. 95: Quasi One-Dimensional Conductors I. Proceedings, 1978. Edited by S. Barišić, A. Bjeliš, J. R. Cooper and B. Leontič. X, 371 pages. 1979.

Vol. 96: Quasi One-Dimensional Conductors II. Proceedings 1978. Edited by S. Barišić, A. Bjeliš, J. R. Cooper and B. Leontić. XII, 461 pages. 1979.

Vol. 97: Hughston, Twistors and Particles. VIII, 153 pages. 1979.